D0149263

DIGGING DINOSAURS

Digging Dinosaurs

John R. Horner
and James Gorman

Illustrated by
Donna Braginetz and Kris Ellingsen

Workman Publishing, New York

Published simultaneously in Canada by Saunders of Toronto Limited.

Library of Congress Cataloging-in-Publication Data

Horner, John R.
Digging dinosaurs.
Includes index.
1. Dinosaurs—Montana—Willow Creek Region (Teton County)
2. Paleontology—Montana—Willow Creek Region (Teton County)—Collection and preservation.
I. Gorman, James, 1949- II. Title.
QE862.DSH643 1988 567.9′1′0978655 88-40226 ISBN 0-89480-220-8

Design by Susan Aronson Stirling
Cover photograph by Terry Husebye

Workman Publishing Company
708 Broadway
New York, New York 10003

Manufactured in the United States of America

First printing October 1988

ACKNOWLEDGMENTS

THIS BOOK IS DEDICATED to Bob Makela, a best friend for 20 years. He always managed to invent a way carefully to pry loose from the rocks the specimens or data I needed, and he was devoted to finding new ways to disseminate scientific information to children of all ages. Bob died in 1987 while working in the field, but his encouragement to others and his determination won't be forgotten. He was a truly exceptional man, and his memory remains with everyone who knew him.

There are a lot of other people who contributed to the research described in this book and to the book itself. I would like particularly to thank Marion and John Brandvold and Laurie and David Trexler for their discoveries and their cooperation. I also want to thank: Don Baird for his support of a preparator's research; Jill Peterson for her work in helping to run the dig and study the fossils we excavated there; Armand de Ricqlès for giving very generously of his time and expertise when we worked with him in Paris; Will Gavin, John Lorenz and Jeff Hooker for their work on which parts of the book are based; John and James Peebles and their families for allowing us to dig on their land for six full summers; the Nature Conservancy for buying the area of the Willow Creek anticline dig and preserving it; and, for all their helpful ideas, discussions and support, Bob Bakker, Bill Clemens, Phil Currie, Peter Dodson, Bob Fields, Gene

Gaffney, Mark Goodwin, Mick Hager, Richard Liebmann-Smith, Dale Russell, David Weishampel, François Vuilleumier and Don Winston.

No account of my debts would be complete without recognizing the people who did the digging, my field crews. Most of them were volunteers, and without them there would have been no dig, and no book. To the many who stopped by to visit and help for short times, thanks. And to those who were there for significant chunks of time, the greatest thanks. In a few cases their last names have slipped from memory, but they were:

1978: Amy Luthin and Bob Makela.

1979: Tony Alexander, Jayne Biakowski, Barbara Haulenbeek, Amy Luthin, Bob Makela, Jane and John McHugh, Fran Tannenbaum and Gay Vostreys.

1980: Dave Archibald, Wayne Cancro, Debbie (?), Barbara Haulenbeek, John Horner (my father), Jason Horner, Bob Makela, Jill Peterson and Bob Schock.

1981: Wayne Cancro, Doug Chadwick, Daren Genereau, Walt Goddard, Jason Horner, Bob Makela, Jay Makela and Jill Peterson.

1982: Bill Abler, Bob Downs, Doug Henderson, Jason Horner, Lora (?), Bob Makela, Jill Peterson, Darren Tanke and Nancy Henderson.

1983: Carrie Ancell, Wayne Cancro, August Cruikshanks, Bob Downs, Bob Franz, Will Gavin, Daren Genereau, Barbara Haulenbeek, Jeff and Diana Hooker, Jason Horner, Vicki Jordan, Robert Leonard, Pam Longibardi, Bob Makela, Dan Martin, Mike (?), Jill Peterson, Carolyn and Sue Purcell, Kate Schank, Bea Taylor and Zoe (?).

1984: Carrie Ancell, Anna Beck-Friis, Bob Downs, George Furness, Will Gavin, Chuck Hannon, Barbara Haulenbeek, Jason Horner, Ken Karzminski, Robert Leonard, Pam Longibardi, Bob Makela, Pat Murphy, Jill Peterson, Phil Peterson, Chris Pinet, Betty Quinn, Walt Spangenberg, Bea and Betsy Taylor, Lisa Ulberg, Robin Voges and Jeri Walton.

The National Science Foundation, the MacArthur Foundation, Princeton University and Montana State University all contributed considerable funding that kept the dig going, for which I am grateful. I would also like to thank my co-author, James Gorman, without whose skill and dedication this story of scientific research would never have become a book. My thanks also to the illustrators, Donna Braginetz and Kris Ellingsen, to Ken Carpenter and Matt Smith, and to Sally Kovalchick and Lynn Strong of Workman Publishing.

Finally, my heartfelt thanks to Mom, Dad, Jim and Rosemary, my son Jason and my wife Joann.

FOREWORD

How can one account for the thrill of finding a fossil? Partly it comes from the straightforward excitement of unearthing buried treasure; partly, from the romance of realizing that the object in your hand was alive millions of years before mankind appeared on earth; and partly, from the exultant realization that, no matter how common the fossil you have found, you are the first human being to see that particular one.

Oddly, few books about fossils convey anything of this delight. *Digging Dinosaurs* is the exception. Here, unusually and thrillingly, is captured all the excitement of the search and the discovery. I have had the rare good fortune of crawling alongside Jack Horner as he made his way up the side of a gulch in the Montana badlands, picking out, with his uncannily sharp eye, fragments of dinosaur eggshells and the tiny bones of the nestlings that emerged from them, and of hearing him bring to life in words the scene around us when it was thronged with vast numbers of nesting giants. I can, therefore, vouch for the accuracy with which this book re-creates that experience—and that, by itself, will make it a joy to read for anyone who has any interest in the natural world.

But this book is something more. Lots of us find fossils. A few people find new species of fossils. But only one or two have the insight and deductive skill, the persistence and sheer good fortune to make discoveries that lead paleontology into completely new areas of interpretation. Jack Horner is one of those people. Not only has he discovered new kinds of dinosaurs, he has revealed whole new aspects of their behavior that bring them to life as never before.

His account of how he did so is a kind of detective story. It begins with the discovery of clues and ends by using them to solve a mystery. Like all good detective stories, it is difficult to put down and you pant to know what will happen next. But there is, of course, a huge difference. Detective stories unravel the mystery of a single fictional death. This story reveals the truth about a multitude of actual lives. No extinct animals have gripped our imagination more vividly than dinosaurs. Few have been written about more extensively. But this marvelous yet modest book sets up a new milestone in the advance of our understanding of these astounding creatures.

David Attenborough
London, August 1988

CONTENTS

INTRODUCTION

If you drive through Choteau, Montana, going south, and take a right just past the Triangle Meat Packing plant, you find yourself on a good gravel road. Go up that road about 10 miles, well past the missile silo, and you come to a rutted track that heads off into scrub pasture—land that looks like nothing but low grass, eroding hills and harsh gullies. About a square mile of this territory was bought recently by the Nature Conservancy, an organization known for preserving special habitats of rare plants and animals. The Nature Conservancy owns, for instance, a nearby swamp (very rare in the plains of eastern and central Montana) called the Pine Butte Nature Reserve, which has probably the highest concentration of grizzly bears per acre of any spot in the United States. But this other bit of land was not bought to preserve bears, or other living animals. The Nature Conservancy acquired it to preserve something else entirely, fossils of animals that are already extinct—dinosaurs.

I first saw this bit of pasture in July of 1981, when I went to

Montana to interview a paleontologist named Jack Horner. The square mile that the Nature Conservancy bought was then part of the Peebles ranch. It was marked on maps of the U.S. Geological Survey as the Willow Creek anticline. Jack Horner was a preparator from Princeton University who had struck it rich, paleontologically speaking, in this spot of ground. He had been digging there since 1978 and, by the time I arrived, had already made quite a name for himself. The discoveries from the anticline were unprecedented. He had dug up fossils of baby dinosaurs and their nests, and he claimed, to the delight of some and the skepticism of others, that these dinosaurs were not only making nests but taking care of their young, just as if they were not dinosaurs at all but immense, leathery robin redbreasts.

Jack Horner also had a reputation for being, in a modern world of scientists for whom the Ph.D. is an absolute prerequisite, something of an eccentric. He had been to college and graduate school; he had a good education in paleontology. But he had neglected to pick up any degrees, or, if you want to look at it from another angle, the University of Montana had neglected to give them to him for reasons having primarily to do with the French language and certain other unfulfilled academic requirements. Jack knew his geology and his vertebrate paleontology, and he knew how to find and excavate fossils, but as far as credentials went the best he could do then (he has since gotten an honorary doctorate from the University of Montana) was a high school diploma.

I had flown to Great Falls, Montana, from New York the day before. At the time I lived in Manhattan, and instead of driving straight out to the dig I thought I needed a short period of decompression to adjust to my new surroundings. I spent the evening at the rodeo at the Montana State Fair. The next morning, feeling no less an Easterner than I had the night before, I drove for an hour or so through cattle range and fields of wheat to get to Choteau. It was my first experience of the plains, and the colors made the strongest impression on me. The

East is green. The plains, at least there, at the foot of the Rocky Mountains, are brown. The prairie grass is brown, the wheat is brown, the horses are brown, and they are all set off by the "Big Sky" that Montana touts on its automobile license plates. The sky lives up to its cliché. It's the dominant feature of the landscape. You live, when you live on the plains, with a 360-degree horizon marking the edge of what seems to be a giant blue-and-white bowl that has been set down on top of you like a walnut shell over a pea.

I had first met Jack in much different surroundings the winter before at Princeton University. He had seemed a bit out of place there. Tall and slightly stoop-shouldered, he looked as if the Gothic architecture of the campus and the close horizon of the East were pressing in on him. As he himself felt they were. He never really liked living in the East. Even the eastern forests left him feeling a bit hemmed in, because, unlike the plains, they don't let you see where you're going. Besides, in the East there's so little relief, in two senses. There's very little relief from the constant presence of buildings, people and expressways. And there's nowhere near enough *geological* relief, which is, roughly, the distance between the high points of the landscape and the low points. Jack had grown up at the western edge of the plains, with the mountains in the distance. He had gone to school in Missoula, on the other side of the continental divide, and had come to relish the enormous, harshly abrupt change the Rockies brought to the open quiet of the plains.

I wouldn't say Jack is shy, that's not the right word, but he isn't always easy to talk to. Sometimes he gets going on the subject of dinosaurs, and he's quite voluble. Sometimes he's silent, for a long time. Sometimes you ask him a question and he thinks about it for longer than I'm usually able to wait for an answer. It's not that he doesn't like to talk, or that he's unfriendly; I've often seen him charm an audience, large or small. I've thought about this quite a bit, during the several years that we struggled with this book, and what I've decided is that his conversation is paced by an internal clock. Sometimes it runs fast, and sometimes it runs slow, but it always runs at its own speed. If you're talking to him and you want to find out what he has to say, you have to adjust yourself to that clock.

Jack not only talks at his own pace; he goes in his own direction. You might say he has an internal compass as a companion to the clock. This is true literally, in that he has a near perfect sense of direction. He always knows where he is and where the cardinal points of the compass are, whether he's in the field or at 79th and Central Park West in Manhattan. When he first came to New York City to do some work in the dinosaur collection at the American Museum of Natural History, he was overwhelmed by the crowds on the street, by the noise, by the size and complexity of the city. Nonetheless he could descend into a subway, take it halfway up or down the island, come out again into plain air and know immediately, without thinking, which way was north. I lived in Manhattan for almost 10 years, and I was never able to do that.

Jack also has his own compass in a personal and professional sense. He has been interested in dinosaurs since he was a child. He has collected, cataloged and kept their fossils since he was seven. And apart from a minor digression to consider astrophysics as a career (it didn't consider him), he always planned to work in paleontology, with degrees or funding if he could get them, without them if he couldn't. I don't mean to paint him as a romantic figure, struggling against all odds to pursue his great love of dinosaurs. I don't see him that way, and he certainly doesn't see himself that way. I think it's simpler than that. Dinosaurs are what interest him, and he pursues what interests him, not what interests other people. As it happens, he has had great success—the result of talent, hard work, and luck. And he relishes that success. But I think no one who knows him has any doubt that he would be doing exactly the same thing (albeit with considerably smaller grants) whether or not his picture was in the newspaper.

Paleontology may seem from the outside to be a simple, very

concrete science. It is the study of the history of life on earth, of organisms that lived once and are now extinct. The only way to study these creatures is to exhume remnants of them that we find somehow preserved, say in rock, or peat bogs, or amber. Vertebrate paleontology, which is Jack's field, is concerned with the fossils of vertebrate animals, such as *Tyrannosaurus rex* and *Stegosaurus*. Yet paleontology exists not primarily in the fossil bones but in the thinking of the paleontologists who collect, organize and interpret these bones. The scientists who study dinosaurs have to construct, within the restraints of theory and evidence, an image of what the dinosaurs were. A bucket of fossils means next to nothing to me or to other laymen. The parts don't come close to adding up to a whole. The paleontologist has to be there, not only to know which bits to gather and how deep to dig, but also to bring his scientific imagination to bear on those bits and pieces in order to vivify them.

The first thing Jack did when I arrived at camp on that first trip was to sit me down and give me a beer. This is always the first thing Jack does with anyone, at camp or anywhere else. The camp was just what it sounds like—three teepees, a lot of scattered small tents, a kitchen (a couple of picnic tables, a gas refrigerator, coolers, and camp stoves under a tarpaulin), and cars and trucks here and there. There were odd bits of paraphernalia about, rock hammers, maps, ice picks, whisk brooms and plaster of Paris. It looked like a group of scientific hoboes had found a place to hide until the police came to roust them out. I met the other members of the dig who were there: Bob Makela and Jill Peterson, who helped Jack run the camp, and a collection of 15 or so volunteers who did most of the hard physical work on the dig. Then Jack and I walked around the anticline. What we walked over was about a square mile of short grass and crumbling dirt and rock. It was eroded into small hills, gullies and ridges. There were mudflats, which looked like dry hardpan soil. But only the surface had dried, not what was under the surface. If you walked on these flats, you could sink to your knees and you'd probably need somebody with a winch to get you out. In a few places pits had been dug, shallow disturbances in the earth where volunteers were shoveling or kneeling to brush dirt off black chunks of fossilized dinosaur bone. I had to keep telling myself that these unimpressive bits of rocklike rubble were what had so impressed other paleontologists.

It took some time before I began to see the detail and complexity in the landscape—the wild honeysuckle and prickly pear, ground squirrels and black widow holes, badger dens and dinosaur fossils. And it took at least as much time talking to Jack, and listening, to begin to penetrate the nature of the scientific activity going on there and the nature of the animals that had once lived in that place. It was a slow process, but I got hooked on dinosaurs on that first trip to the dig. And I ended up listening to Jack, and following him around, from Montana to Paris and back to Montana, during the three years it took us to write this book. With the exception of some new discoveries reported in the last chapter, everything described in the book came from two square miles of cattle range over the course of seven years.

The finds came at a time when the dinosaurs had been extinct for 65 million years and paleontologists had been studying them for 150 years. In the 1970s, however, the world of dinosaurs had been turned upside down. The old image of dinosaurs had been one of coldblooded, slow, dumb reptiles. Well, they were reptiles all right, but suddenly paleontologists were saying they were warmblooded, they were fast, they were smart. They didn't move the way we thought they did. They stood differently. In short, we needed to reimagine them.

But by and large these debates were conducted on the basis of the old evidence. There were a few new finds, but not startling ones. Then, to cap off a decade of argument and ferment, paleontology was offered the Willow Creek anticline. Here was new evidence, a lot of it, and amazing evidence—dinosaurs feeding their young (in the nest), dinosaurs nesting in colonies, dinosaurs in gigantic herds. And Jack kept finding, indeed is still finding, more babies, more nests, more remarkable fossils. As the book was going to press, he had just found a fossilized nest of dinosaur eggs with the babies preserved as they were in the process of hatching.

What I tried to do, in my part of our collaboration on the book, was to recapture the experience I had when I first went out to the dig in 1981. I remember seeing the fossils, seeing spray-painted orange circles that marked where nests had been dug up, and thinking to myself that 80 million years ago little dinosaurs had been bouncing around in those little circles. But most of all, I remember the excitement of having Jack bring the scene to life. We sat, one afternoon, on a hillside from which we could look at all the different sites being excavated. All I could see was scruffy hills and exposed rock. But I kept my mouth shut, and at his own pace Jack described to me what everything looked like when the dinosaurs were alive. To him the land as it had been 80 million years ago, in the late Cretaceous, was at least as vivid as the way it was at the moment. He pointed here and there as if he could see what he was describing. Over there was a nesting ground with mother dinosaurs running back and forth from it to the stream. The stream ran right there, through that block of green mudstone. And here, that was a lake, and right out into it jutted a peninsula or perhaps it was an island. On the island, probably hidden by tall sedge, waving in the breeze, were small dinosaurs, laying eggs. Maybe later in the day other dinosaurs would be stealing those eggs. Over the lake flew pterosaurs. Bit by bit, Jack lifted the skin of dirt and vegetation, the accumulated dust of 80 million years, and showed me what was underneath it. What he showed me, in all the detail that paleontology can muster, was the lives of the dinosaurs. If you listen, I think he'll do the same for you.

James Gorman
New York, January 1988

CHAPTER 1

LOOKING FOR BABIES

In the winter of 1978, I was working as a preparator at Princeton University. Preparators aren't professors or curators, and under normal circumstances they have next to no chance of becoming professors or curators. Usually you find them in the dusty, windowless basements that more often than not pass for laboratories in paleontology. In the summer they do go along on expeditions, hunting for fossils in the badlands of Montana, or Colorado, or Mongolia. But then, come September, somebody has to clean up the discoveries and make them look interesting. That's what preparators do. Confronted by chunks of rock with bits of fossil bone in them, or fragmented fossil bones, they chip away the rock or dissolve it with acid to liberate the fossils. Then they try to put the fragmented and broken fossil bones together. The result of their work may be something as small (but important) as the rebuilt jaw of a new variety of dinosaur, or as large and impressive as a full skeleton of *Tyrannosaurus rex* in a museum display.

A preparator's lot can vary greatly depending on his boss. He may be only a backroom technician, a hired hand with no say in the research. Or he may work as a junior partner in the research, collaborating with his boss rather than just scrubbing clean the summer's haul of femurs and tibias. My situation was probably the best a preparator could hope for. My boss, Don Baird, and I worked together on several research projects. And it was Don who introduced me to the fossil collections of the great museums. Traveling up and down the East Coast, we inspected the collections at the American Museum of Natural History in New York, the Philadelphia Academy of Natural Sciences, the Smithsonian Institution, and the Carnegie Museum in Pittsburgh. We spent our time not so much in the public halls, where the impressive skeletons are displayed, but in the back rooms and basements, looking at shelves of fossil bones that had been dug up years before and were kept there as a resource for paleontologists to study.

We examined these old fossils partly to pursue research interests of Don's and partly to continue my education. Before 1975, when I arrived at Princeton, my career in paleontology had been somewhat checkered. I'd been collecting dinosaur bones since I was seven years old, and I'd taken every undergraduate and graduate course in biology and geology that the University of Montana had to offer, but academically I'd had a little trouble holding to the straight and narrow. I'd ignored a few of the humanities and had never fulfilled the degree requirements. For a while, after leaving the University of Montana, I stayed in Shelby, my hometown, running the family sand-and-gravel plant with my brother. But crushing rocks held no interest for me. It paid better than paleontology, but I couldn't manage to corral the enthusiasm I had for dinosaurs and apply it to this way of making a living. I wrote letters to all the natural history museums in the English-speaking world, twice, inquiring about work. I also haunted the meetings of the Society of Vertebrate Paleontology, which serve as a job market (such as there is) in the field, and through this process I found the job at Princeton.

Don also encouraged me to do my own research. Until the winter of 1978, I'd had no particular interest in baby dinosaurs. I was interested in what are commonly called the duckbills. I'd done quite a bit of prospecting for dinosaur fossils in Montana and Alberta, Canada, on my own and with my friend Bob Makela. Duckbills were what I found, so duckbills were what I looked for. In some ways, scientific research is like taking a tangled ball of twine and trying to unravel it. You look for loose ends. When you find one, you tug on it to see if it leads to the heart of the tangle. Sometimes the loose end leads nowhere; sometimes it leads you deeper into the ball, to unexpected and intriguing knots. I guess you could say that dinosaurs were my ball of string. Duckbills were the only loose end that I had been able to find, so I had been tugging on them for all they were worth.

In our own collection at the small Princeton museum, Don and I found some duckbill specimens that had been overlooked in the scientific literature. They had been collected around 1900 by Earl Douglass in Montana in a rock deposit called the Bear Paw shale. This was a marine sediment, meaning that the rock was formed from the bottom muck of a sea. Fossils found there came from animals that had died and sunk into the muck millions of years ago. Some of these animals were sea creatures, but not all of them. The dinosaurs, at least, were land animals. There may have been dinosaurs that paddled about in lagoons, looking for aquatic vegetation, but no dinosaur was seagoing. The plesiosaurs and other great marine lizards that swam in the seas were an entirely different sort of reptile.

The dinosaurs whose bones Douglass found in the Bear Paw shale might have died on a beach or while foraging for food in shallow water; in this case, they would have been washed out to sea and their bodies would have settled into the bottom muck. Or they might have died farther inland and their bones might have been washed to the sea

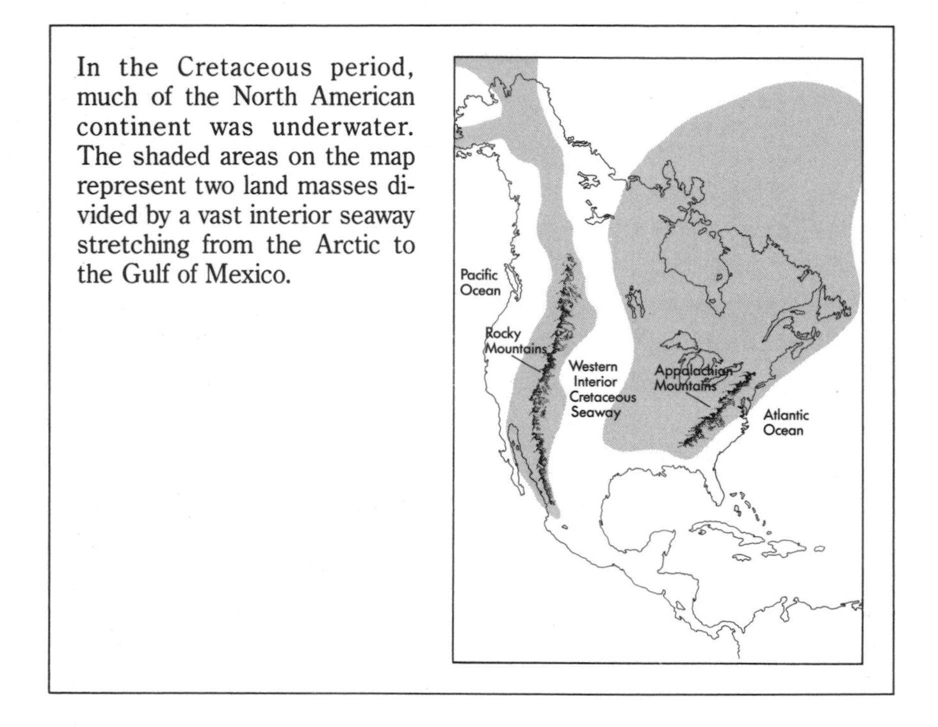

In the Cretaceous period, much of the North American continent was underwater. The shaded areas on the map represent two land masses divided by a vast interior seaway stretching from the Arctic to the Gulf of Mexico.

by rivers and streams. This sea no longer exists, but at the time these dinosaurs lived, a little more than 70 million years ago, it covered the center of North America between the then emerging Rocky Mountains and the already old, eroding Appalachians. Geologists call it the Western Interior Cretaceous Seaway. The dinosaurs that are preserved in the Bear Paw shale spent their lives on a coastal plain, at its largest something like 200 miles wide, between this seaway and the Rockies.

In New Jersey and Delaware, fossils of dinosaurs had also been found in marine sediments. (In fact, this was the origin of all the dinosaur fossils I had seen from these two states.) These sediments had been deposited not by the inland sea but by the Atlantic Ocean, which, millions of years ago, covered much of the land that now makes up New Jersey and Delaware. The dinosaurs preserved here had lived on the coastal plain between the Appalachians and the Atlantic.

I became curious about dinosaur fossils from marine sediments. What did they have in common? Did they share any traits that would explain why they had been found in these old sea bottoms? I made a catalog of all the known dinosaur fossils from marine sediments and looked for similarities. What I discovered was that most of them were duck-billed dinosaurs, which was not particularly informative, and that 50 percent of the animals found in marine sediments were juveniles. That latter bit of information was no small discovery. Of all dinosaur fossils, those of young dinosaurs were the rarest of all.

IT'S HARD TO OVEREMPHASIZE how scarce juveniles were before 1978. You saw full-grown dinosaurs striking familiar poses in every natural history museum in the world, but you saw very few young. Over the 150 years that paleontologists had been hunting and finding dinosaur bones, fossils of juveniles had been found so rarely that their scarcity had become a major scientific puzzle. Other than one spectacular find in Mongolia, which I'll come to shortly, there had been some finds in New Mexico, isolated bones in France, footprints in the Peace River in British Columbia, some misidentified juveniles found in Montana and neighboring Alberta, and a few others here and there.[1] If I were to make a simple list of fossils of adult dinosaurs that had been found, it would fill a book-length catalog, probably several such catalogs. The fossils of juveniles could be listed in a short pamphlet.

The fossils of *baby* dinosaurs could have been put on an index card. In 1978 these were all the known fossils of baby dinosaurs: some coelurosaurs from the Ghost Ranch in New Mexico; footprints of baby duckbills in shale in a coal mine in Utah; a dozen or two isolated bones of babies found in Canada and Montana; a baby found in the Judith River formation in Canada and described in 1956; a well-preserved baby skeleton from Argentina; and one tremendously impressive collection of fossils from the Mongolia find that I mentioned, including adults, eggs, and juveniles of all ages from hatchling on up. Finds of dinosaur eggs were just as rare. There were those from Mongolia, quite a number from a different sort of dinosaur in the south of France, isolated finds from India and Africa, and eggshell fragments from North and South America.[2]

The Mongolian group, on display at the American Museum of Natural History in New York City, constituted the one and only major find of baby dinosaurs and eggs in the history of paleontology. It was a fantastic discovery. The fossils were found by the American Museum's Third Asiatic expedition, which had gone to Mongolia in 1922 to look for the origin of man. During the trip home across the desert, when the jeep caravan stopped to rest, a photographer took a walk and stumbled upon what turned out to be a dinosaur skull and eggshell. The next year the expedition returned to mine the site fully and found the first dinosaur eggs the world had seen. Until then scientists had suspected that dinosaurs (like many other reptiles) reproduced by laying eggs, but they had never had the evidence to prove this point of view. Here that evidence was—in abundance. Fifty eggs were discovered, as well as countless fragments of eggshell. The eggs were found both as individual specimens, lying on the ground, and in one of about four clutches, or nests. In that one single season in Mongolia, the expedition dug up, in addition to the eggs, several skeletons and more than 50 skulls of *Protoceratops andrewsi,* the dinosaur that laid and hatched from the eggs. The skulls were of animals ranging in age from hatchling to adult, with all stages in between. The researchers ceased work after five weeks of digging, even though they were still discovering new specimens. They carried home 60 cases of fossils packed in camel hair and weighing 10,000 pounds.[3]

These Mongolian fossils came from red sandstone, the preserved remnants of a dry coastal plain. And their discovery raised the question of why these so-called "red beds" were chock-full of baby dinosaurs when these fossils were scarce everywhere else. As succeeding years yielded no other major finds of baby dinosaurs, the question grew in importance. If you think about it (as I began to do once I had noticed the presence of juvenile fossils in the Bear Paw shale), more dinosaurs should have died young than died old; that's what happens with most animals. And that high infant mortality should have produced a lot of fossils over the course of 140 million years—a lot of fossils that had never been found.

Several explanations had been offered for the scarcity of fossil remnants of juvenile dinosaurs. Perhaps the bones of the young were not as sturdy and thus did not become fossils as easily as the bigger, harder adult bones. But, if that were the case, why had paleontologists frequently found fossilized bones of lizards and other small reptiles? Another suggestion was that the young lived in a different area from the adults, the way hatchling sea turtles disappear for a year. It had also been suggested that, to lay eggs, adult dinosaurs might have migrated to drier areas, those sections of coastal plain farthest from the various oceans on whose borders we know the dinosaurs lived. The Mongolian red beds were just such an upper coastal plain.[4]

If the dinosaurs laid their eggs primarily in the upper sections of coastal plains, then it would make perfect sense that so few eggs, babies and older juveniles had been found. As a coastal plain reaches farther from the sea toward the mountains, the pitch of the land increases so that the upper sections are constantly being stripped of soil by wind and water. The steeper pitch means that the streams run faster, and dust and dirt are pulled, by gravity, inexorably downward. In this situation, the bones of animals that die are likely to be washed or blown away. Even when the streams that run through these upper coastal plains do leave deposits of soil or silt, the accumulation is not as great as it is in the lower coastal plains and not as effective in preserving fossils.

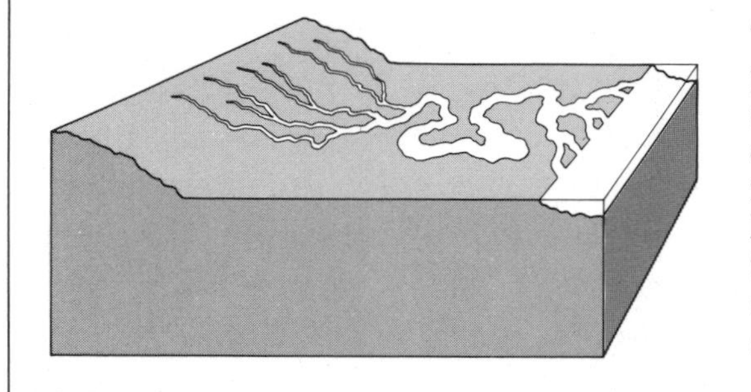

The upper coastal plains (*at left*) were subject to erosion caused by small, fast streams; in the lower plains (*right*), the deposition from slow rivers built up silty deltas.

The lower sections of a coastal plain, nearer whatever sea the plain borders, are flatter. Here one finds slow, meandering streams and rivers, which deposit heavy burdens of silt on their banks during floods. They may also build up large deltas. Deposition is always occurring, and bones are thus likely to be buried and preserved. In consequence, modern paleontologists usually explore the preserved remains of ancient lower coastal plains, where fossil deposits are predictably rich. From the 1920s to the 1970s, plenty of fossils of adult dinosaurs were found in these lowland deposits. On the infrequent occasions when paleontologists did explore preserved upper coastal plains, they sometimes did find fossils of young dinosaurs but in many cases failed to recognize them for what they were.

The Mongolian red beds were the exception. The American Museum paleontologists suggested that the reason for the abundance of well-preserved fossils in this upper coastal plain was that the sandstones there were aeolian, or wind, deposits. That is to say, this was a dry, almost desert-like area when the dinosaurs lived there and the winds carried fine sand and deposited it much the way a river carries and deposits silt.

It was in the context of this dearth of juvenile dinosaurs that I came upon the preponderance of such fossils in Douglass' finds from the Bear Paw shale. I had no idea, in the winter of 1978, why there should be so many in that shale, why nobody had noticed the numbers before, or why other shale deposits hadn't yielded similar numbers. All I knew was that the shale contained juvenile fossils, perhaps even babies, that baby dinosaurs were a coveted rarity—and that if I got myself out to Montana the next summer to take another look, I might find some baby dinosaurs myself. So, to prepare for the search, I inspected the museum collections in the East and studied all the fossils of young dinosaurs I could locate. I wanted to have a clear idea in my mind of what, precisely, I would be looking for when I got out into the field.

MY FIELD SEASON THAT YEAR was my vacation. In 1978 I had neither the position nor the funding to go off hunting dinosaurs for the whole summer at the university's expense, the way an established, degreed paleontologist would. I did, however, have four weeks off in July.

I went straight to Rudyard, Montana, where Bob Makela was waiting for me. Rudyard is on Montana's high line, the string of towns near the Canadian border, about 150 miles from the Rocky Mountains. It's not far from Shelby, where I grew up. Farming and ranching country spreads out all around these towns, flat and sparsely populated as most of Montana is east of the Rockies. You drive for miles between towns, and most of them are just a flicker of buildings as you drive through. The land is dry, and it's usually farmed that way, without irrigation. It produces wheat and cattle, and yields numerous fossils of Cretaceous shellfish.

Bob had been teaching high school science in Rudyard since he graduated from the University of Montana, where we first met. He and I were both interested in dinosaurs, and we'd both gone out on university-sponsored field trips in geology and paleontology. We became good friends. Bob was a remarkably forceful man. He was energetic, outgoing and optimistic. He loved hard work and nothing intimidated him. In fact, some things that should have intimidated him didn't. The first time I saw him, he walked into a herpetology class wearing a T-shirt and cradling a gila monster in his arms. Gila monsters are quite poisonous, and most people don't carry them around unless they've got some kind of protective clothing—gloves, or a heavy shirt at the least. But Bob didn't seem to mind the gila monster, and the gila monster didn't seem to mind him.

I don't think of myself as a milquetoast when it comes to reptiles, but Bob far surpassed me in that department. He worked in the herpetology laboratory, and sometimes when Carolina, the seven-foot Eastern diamondback rattlesnake, got loose in the lab, he'd go in and poke around under the desks and chairs until he found her and caught her. I gave him moral support. (The other thing that escaped in that lab was a bunch of baby black widow spiders. When they hatched, they were too small to be deterred by the screen that kept their mother in her cage. Sometimes the herpetology professor, who had his office in the lab, would be sitting at his typewriter and across the paper, on all eight legs, would walk a baby black widow.)

Bob and I began collecting dinosaur fossils in the mid-sixties, and we kept at it. I think we were ideally suited to each other. Both of us treated the explorations as an adventure; both of us were willing to do it on our own, whether or not we had the support of a university; and neither of us minded the physical demands of hiking, digging and hauling. In later years, when we had a big dig going, Bob's knack for dealing with people balanced my reticence. He was the person the volunteers on a dig would gather around naturally in the evening to listen to stories of Montana, or dinosaurs, or grizzly bears—whatever stories Bob wanted to tell. I think Bob loved conversation as much as he loved paleontology. It was his teepee that had the fire in the middle where everybody would get together on a cold night for beers. I was usually there, but I wasn't the host. Bob was the social, emotional and

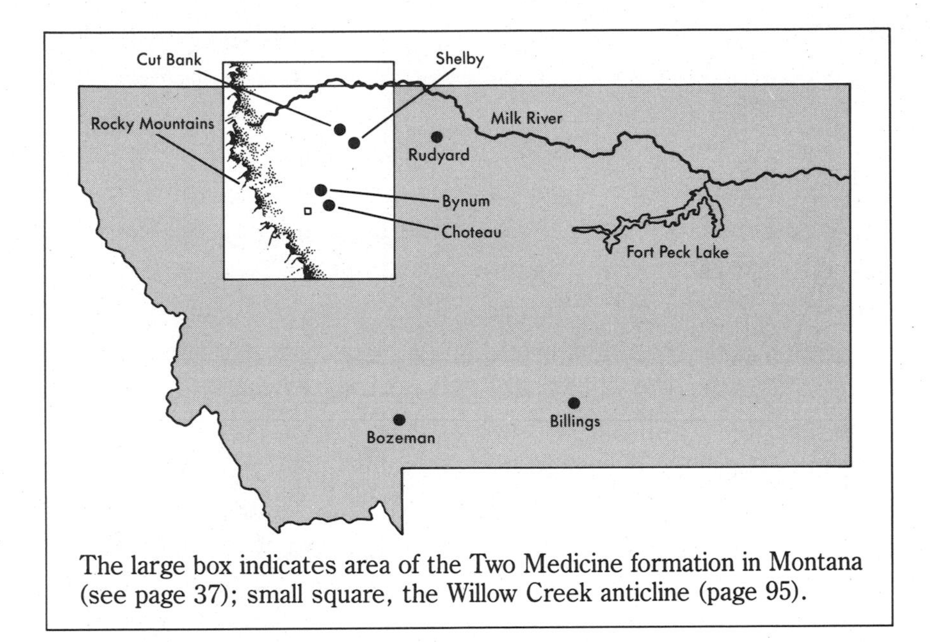

The large box indicates area of the Two Medicine formation in Montana (see page 37); small square, the Willow Creek anticline (page 95).

organizational anchor of the dig. If I provided the scientific direction, the paleontological planning, what you might call the head, Bob provided the heart.

From Rudyard, we drove southeast through the wheat fields and cattle ranches. We were headed to the central part of the state, near Billings. The land there is also flat. The horizon is wide. And underfoot is the Bear Paw shale. As I said before, shale is a marine sediment. And we find it in Montana because during the Cretaceous period the center of North America was occupied not by grasslands but by the interior seaway I described earlier. Over millions of years, this sea would rise and fall, expanding and contracting, leaving bottom muck that eventually turned into shale. At its largest, the seaway extended all the way from the Arctic Ocean to the Gulf of Mexico. It was shallow and filled with seagoing lizards, sharks and other fish. It probably looked something like Florida Bay, the island-dotted, biologically rich finger of the Gulf of Mexico that stretches between the southern tip of mainland Florida and the archipelago of the Florida Keys. At the end of the Cretaceous period, 65 million years ago, this sea finally dried up for good, and left behind the old sea bottom that has become the Great Plains.

The last shale deposit left by this seaway in Montana and Alberta is the Bear Paw shale, patches of which are found throughout the area. The patch near Billings was where Earl Douglass had found the juvenile dinosaur fossils. And it was the place where I hoped Bob and I would dig up a baby or two. We might have, too, if the weather had been with us. But when we got to the site, the sun disappeared, the temperature dropped, and it poured rain for three days. Wherever we walked, we sank above our ankles. Shale is an odd rock when it comes to rain. It suffers a kind of identity crisis and turns back to mud. Each time I pulled my foot out of the muck, I brought a pound or two of the old Cretaceous sea bed with it. It was like walking in a vat of warm licorice, and that's no way to look for fossils. We got in the van and drove away, with no dry shoes, and no baby dinosaurs. I hadn't given up. I planned to go back. But I was going to wait for the kind of summer weather more common to the Montana plains—100 degrees with 10 percent humidity.

We ended up at the Milk River badlands near Rudyard. Like a lot of the American West, Montana has plenty of land that's been scoured clean, and in the summer it's practically crawling with paleontologists. You see, the whole of a paleontologist's professional life is connected to either deposition or erosion. You want deposition to preserve the fossils and then erosion to expose them for you. You don't hunt fossils in a lush forest; you want something to have removed the trees, brush, topsoil and a good bit of the rock to get it ready for you.

The reason we went to the Milk River was that Bob and another science teacher, Larry French, had found some mammal fossils there and we wanted to show them to Bill Clemens, a specialist in early mammals. In a sense, being a paleontologist is like being a member of a club, and club members tend to help each other out. We also help each other because in modern science everybody has a narrow area of specialty; if you find something in somebody else's specialty, you send it on to them or show them where you got it from. As it turned out, Bill and his crew (from the University of California at Berkeley) were quite pleased with the find. It proved to be a very rich deposit, and in a kind of turnabout Bill mentioned to us that a rock shop in Bynum had a dinosaur fossil that the owner wanted identified. He was going to do it, but we were the ones who were really interested in dinosaurs and there was always the chance it might be something good. When the time came for us to leave Rudyard and the Milk River to go to work for a few days on yet another paleontological dig farther south in Montana (on this one a friend was collecting fossil fish), we took a detour through Bynum.[5]

Bynum is as small as a town can get before it just becomes somebody's house on the road. It had, in 1978, a gas station/store, a few buildings, and an old faded church that housed a rock shop. It was Sunday when we got there, but the rock shop was open because the owners were Seventh-Day Adventists and their sabbath is Saturday. The shop was like a lot of other little rock shops in the West: cluttered and dusty, with rock samples, gems, geodes and fossils, all for sale. The fossil in question turned out to be common enough. It was part of the backbone of a duck-billed dinosaur. We were in no hurry, so we wandered around the shop, picking out all the fossils that had been misidentified and giving them the correct identification.

The owners of the shop were Marion and John Brandvold, who now have a small tourist museum and gift shop in Choteau, Montana, a much better location. That Sunday morning, Mrs. Brandvold was running the shop. She's a dark-haired, sharp-featured woman given to fringed jackets and other Western garb. She was delighted that we were identifying everything, and when we told her there was no charge for our service, she asked us if we could identify some bones that she had in her house. She and her family had collected them earlier that year, in the spring. She went back to the house and brought out two specimens. They weren't much to look at, just two dusty pieces of gray bone, but it was immediately obvious to me that they were the hip end of a duckbill thighbone and a bit of a rib—except that they were the wrong size. The femur, or thighbone, of a typical duckbill might be four feet long and as thick as a fencepost. The femur that Mrs. Brandvold handed me, if the bone had been whole, would have been the size of my thumb. It was broken, and all that remained was a piece an inch long.

What I had in my hand was a bone from a baby dinosaur, a duckbill—exactly what I wanted, in a place I never expected to find it. And it wasn't the only one. Mrs. Brandvold took us into her house, and there, spread out on a card table, were numerous small bones. The first thing Bob and I noticed was a jawbone about two inches long. Paleontologists love jaws. They have so much detail to them, and you can learn a great deal about a jaw's owner—what he ate, for instance. Jaws are also completely unmistakable. It requires a practiced eye to see what a piece of a femur is, but a jawbone is a jawbone. An adult duckbill jaw would run about three feet long. This one was two inches long. Bob had been skeptical about the femur, but the jaw convinced him.

We told Marion Brandvold how important the fossils looked, and she agreed to give them to us. She filled up a coffee can with them and presented it to us. When we sorted out the bones later, we could tell by counting the legbones that we had the remains of at least four baby duckbills.

That was enough, in itself, to qualify as a big paleontological find. Certainly for a preparator and a high school teacher, it was a terrific find. But, in paleontology as in life, enough is never really enough. There had to be more where those fossils came from, and the

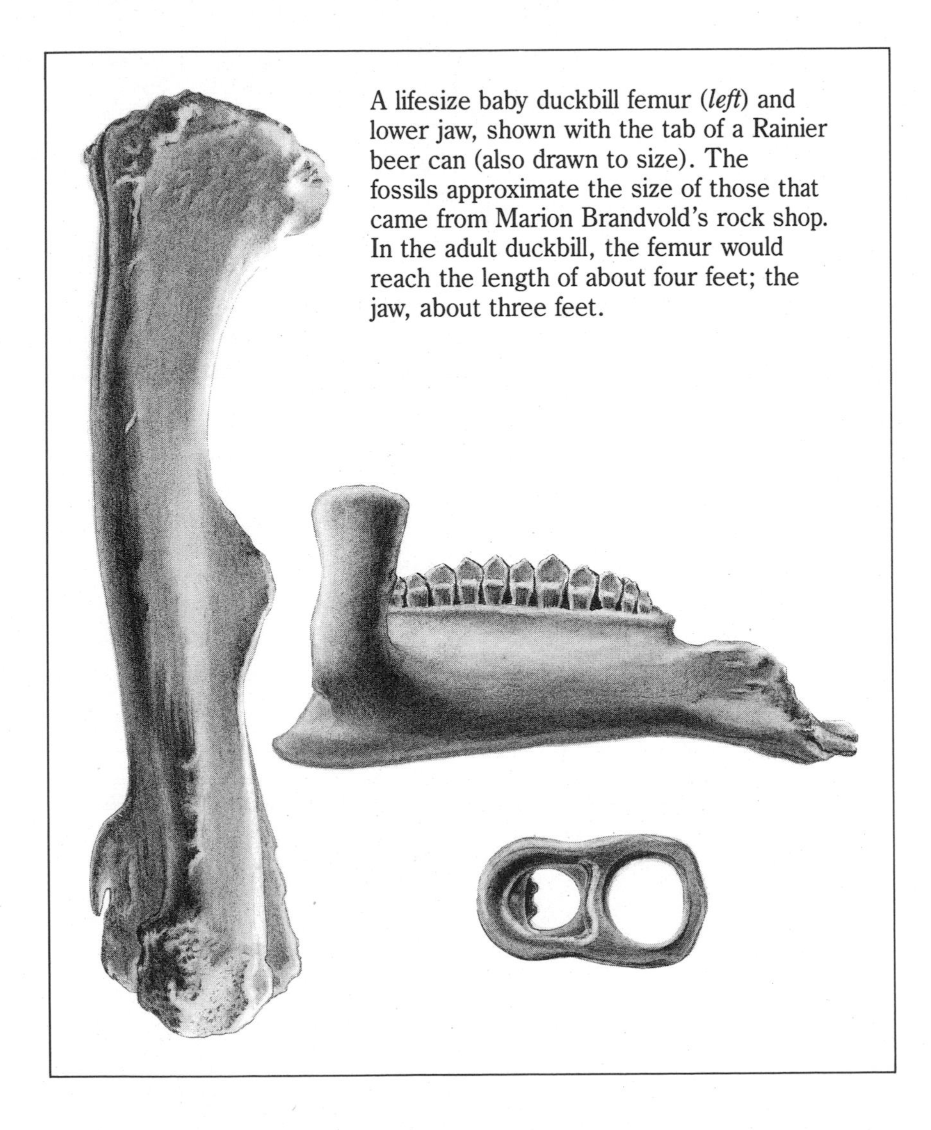

A lifesize baby duckbill femur (*left*) and lower jaw, shown with the tab of a Rainier beer can (also drawn to size). The fossils approximate the size of those that came from Marion Brandvold's rock shop. In the adult duckbill, the femur would reach the length of about four feet; the jaw, about three feet.

question that was foremost in our minds, the question we asked Marion Brandvold as soon as we realized what we were looking at, was: Where *did* they come from? The answer was most definitely not Douglass' site near Billings. Marion Brandvold had found those babies somewhere else, in an area that I knew well but not, apparently, well enough.

THE FOSSILS HAD COME from the Two Medicine formation, a 2,000-foot-thick wedge of sandstones, shales and mudstones stretching over a huge, ragged patch of Montana east of the Rockies.[6] A formation is the basic unit used by geologists to map and catalog the earth's layers of rock. Its vertical and horizontal boundaries are determined by the characteristics of its rock beds. A given formation might, for instance, have an identifiable sequence of shale and sandstone layers derived from a particular sea. For the rock beds to constitute a formation they must, in the nature of the rock itself and in the way they fit together, stand as a unit, apart from their surroundings. There are no size limits on formations, but they are usually pretty large affairs. You could safely say that a formation is bigger than a bread box and smaller than a planet.

The surface of the Two Medicine formation covers 3,600 square miles, running from the Canadian border in the north to Augusta, Montana, in the south and from the Rocky Mountains in the west to the town of Choteau in the east. For the most part the exposed sections of the formation, where one can get at the rock, are on privately owned range land; one big section lies in the Blackfoot Indian reservation just east of Glacier National Park. The surface, however, is perhaps the least important dimension in geology and paleontology. More important is the vertical dimension. In a rock formation, depth is really a measure of time. Vertical feet or meters are a kind of record of the centuries and millennia that had to pass in order for stream sands, sea mud and other sediments to be deposited, buried under other sediments and eventually transformed by pressure into rock. This understanding of sedimentation, that it marks the passage of enormous stretches of time, is necessary for paleontology to exist as a science.

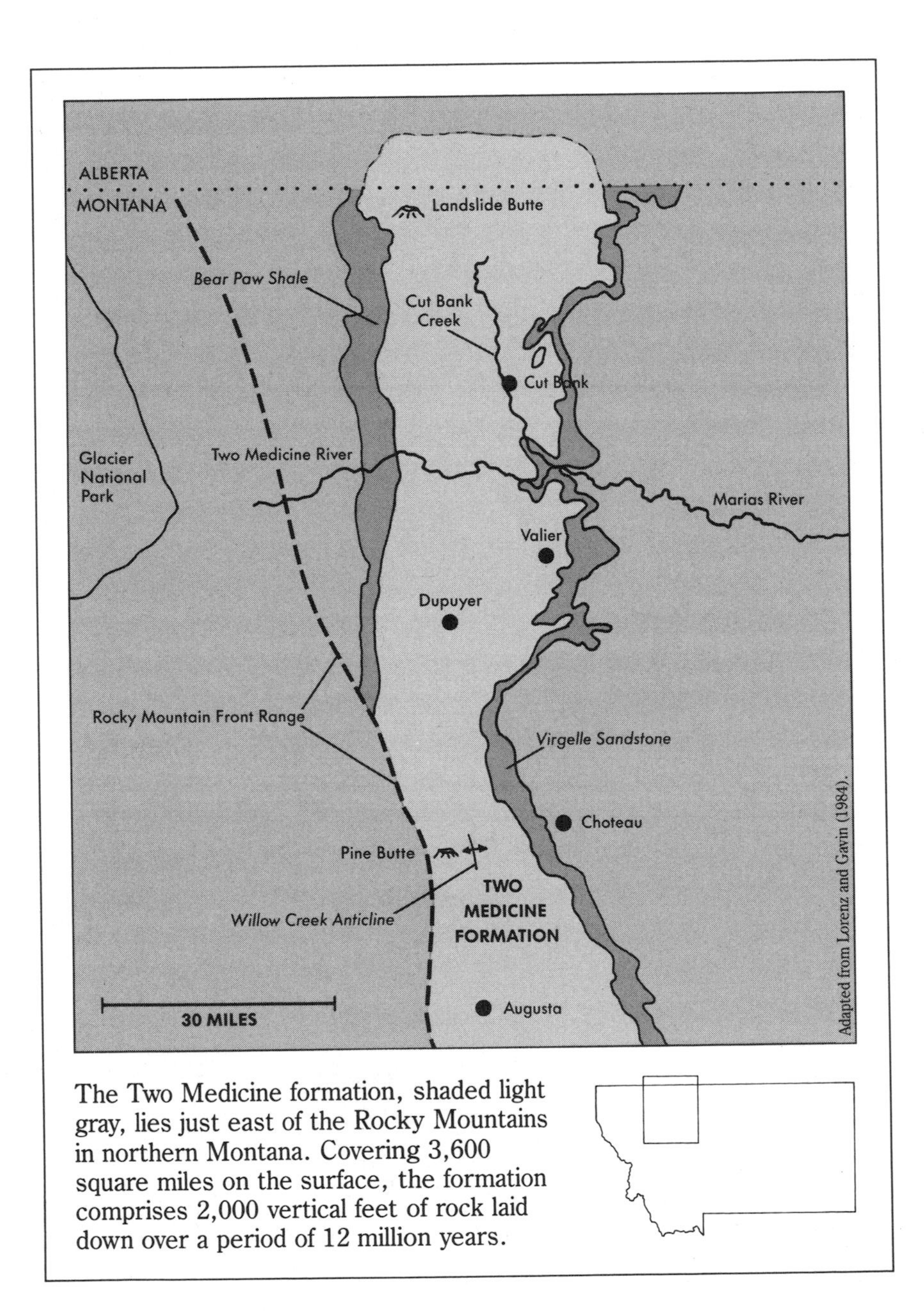

The Two Medicine formation, shaded light gray, lies just east of the Rocky Mountains in northern Montana. Covering 3,600 square miles on the surface, the formation comprises 2,000 vertical feet of rock laid down over a period of 12 million years.

It's also necessary to realize that rock is deposited in chronological sequence. One may have to deal with disturbances, earth movements that bend, twist and overturn perfectly laid down rock layers, but in undisturbed sedimentary rock what's on the bottom is the oldest and what's on the top is the youngest. The 2,000 vertical feet of the Two Medicine formation document the passage of roughly 12 million years, which is a big enough portion in the history of dinosaurs to track the processes of evolution—to watch new species come and go.

The formation is made up of Cretaceous rock. That is to say, the sediments originated in the Cretaceous period, which began about 140 million years ago and ended 65 million years ago. This span of 75 million

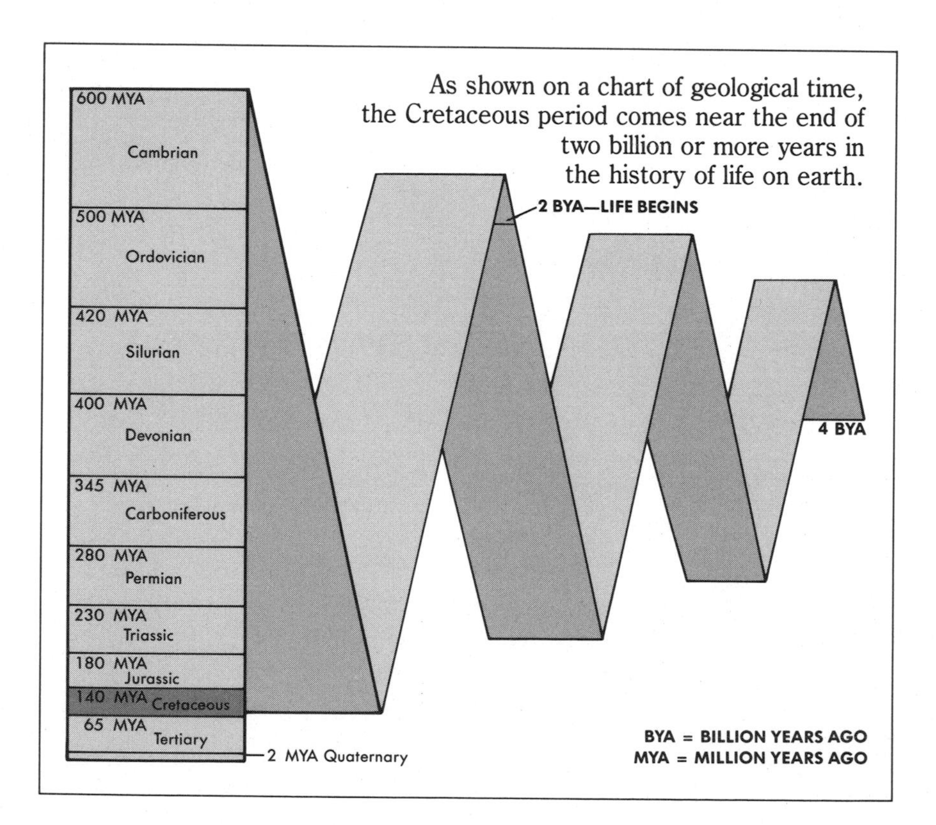

As shown on a chart of geological time, the Cretaceous period comes near the end of two billion or more years in the history of life on earth.

years occurs quite late in the history of the planet, which is probably four and a half billion years old, and late in the history of life, which may have begun more than two billion years ago; it's even fairly late in the history of dinosaurs, which first appeared a little more than 200 million years ago.

The time span of the Two Medicine formation (laid down in the late Cretaceous, from about 84 to 72 million years ago) is best measured not by years, but by the rise and fall of the inland sea. This is the body of water that I described earlier as expanding and contracting. The Two Medicine formation began during a contraction, or recession, that is called, in its northern reaches, the Colorado Sea. It had reached all the way to the Rocky Mountains and had begun to shrink. It's hard to say whether the Colorado Sea actually reached into the mountains and filled valleys there, because we don't know precisely where the mountains were at the time. The Rockies were then very young and in the process of growing, which means there were frequent volcanic eruptions and earthquakes as the planet shuddered and cracked and thrust the mountains up. What we can say is that they were somewhere between where they are now and 50 miles west of that line. Certainly the Colorado Sea went deep into where the mountains are now. Then, as it receded, the sea left deposits of muddy shale and of beach sand that turned into sandstone.

This beach sand deposit, called the Virgelle Sandstone, marks the bottom boundary of the Two Medicine formation. As the sea receded, a long coastal plain opened up—extending from the mountains to the sea, eventually a distance of 200 miles or more. This plain was richly populated by many varieties of dinosaurs. As the mountains thrust and bulged, they shed enormous amounts of dust and rubble that were carried by streams and rivers down to the plain. The streams and rivers overflowed frequently, leaving sand and mud on their floodplains. Over millions of years, such floods deposited enormous amounts of sediment. In addition, the thrusting of the mountains

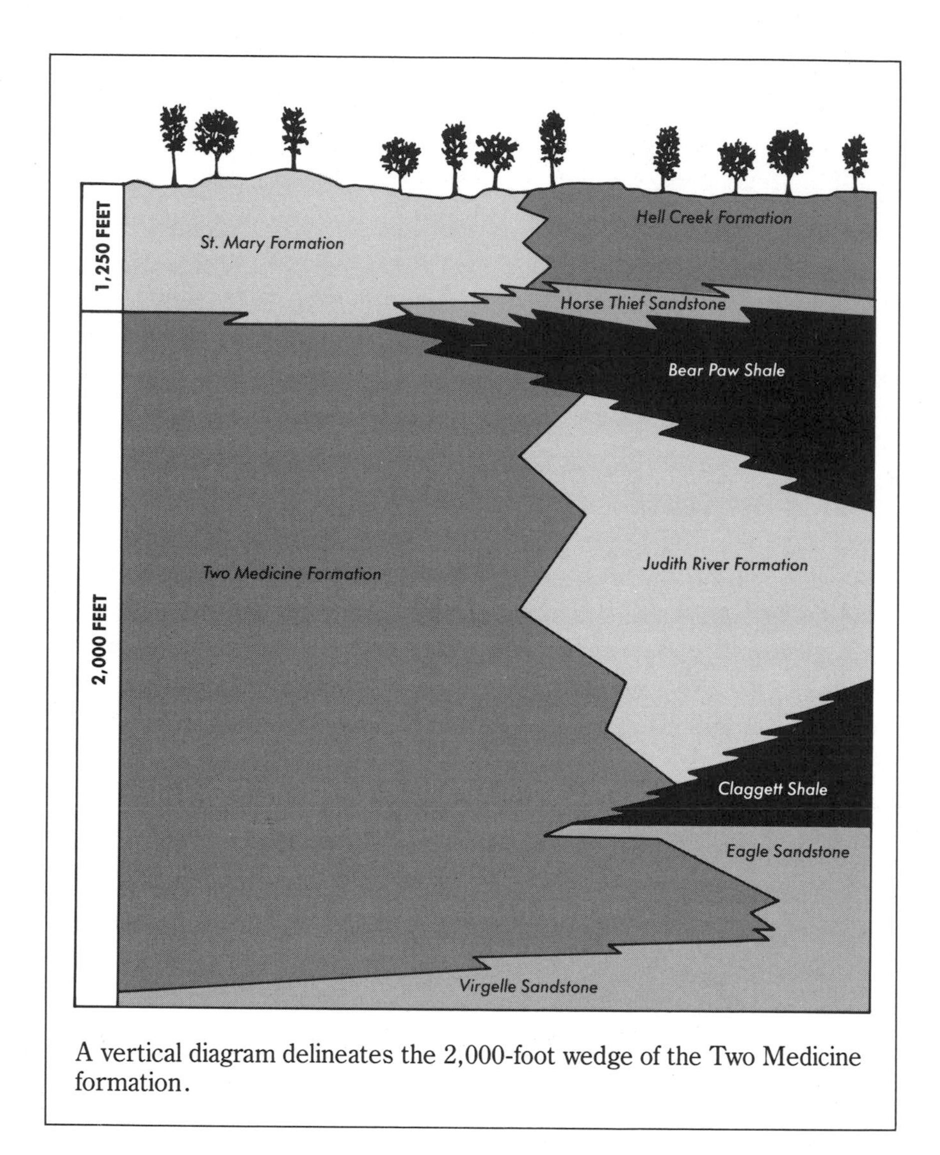

A vertical diagram delineates the 2,000-foot wedge of the Two Medicine formation.

pushed down the land between them and the sea into what is called a foredeep. If, while you're sitting in bed, you scrunch up the blanket to push up a hill, you'll notice a dip in the blanket just on the other side of

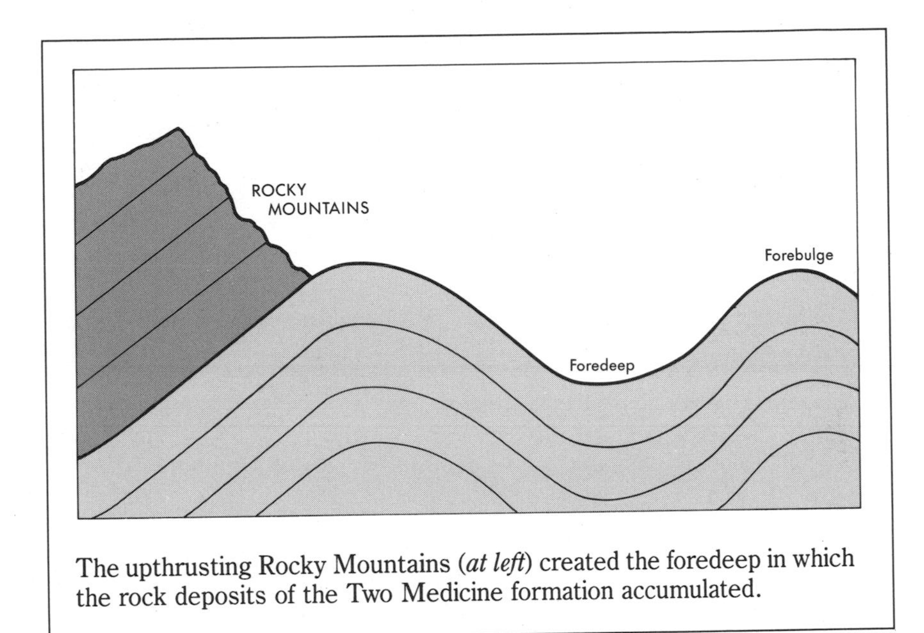

The upthrusting Rocky Mountains (*at left*) created the foredeep in which the rock deposits of the Two Medicine formation accumulated.

the hill. This is more or less what happened on a grander scale with the Rockies and the Two Medicine formation. The Rockies were the hill and in front of them was a dip, or foredeep, that became the Two Medicine formation. Onto this foredeep was deposited first the muck and sand of the receding sea, and then huge amounts of silt and sediment from the rising mountains. With all this sediment, the foredeep sank even more.

Eventually, the sea stopped receding, and after a time it began a new stage of expansion called, in Montana and Alberta, the Bear Paw Sea. As it expanded, the sea inundated the coastal plain. The sea bottom muck fell on top of river and stream silt deposited earlier, and over millions of years that muck turned into the Bear Paw shale. Today the Bear Paw shale does not exist everywhere the sea once existed. Erosion has occurred in some spots, and in other spots the level of the shale is so far beneath the surface of the land that we don't know

In the late Cretaceous, the dry upper coastal plains of western North America
duckbills (*background*) to the large, fearsome *Albertosaurus* (*middle distance*)

were dominated by dinosaurs; from *Euoplocephalus* (*foreground*) and the and fast, graceful *Ornithomimus*, or "bird mimic" (*far distance*).

whether it's still there or not. So today in Montana you find patches of this shale. The portion of the Bear Paw shale that Bob Makela and I had gone to explore, looking for babies, is one such patch. Another patch forms the upper boundary of the Two Medicine formation.

What the Two Medicine formation preserves, then, is the record of 12 million years of life on a coastal plain. During that time the plain grew and shrank as the size of the sea changed, but we know pretty much what the circumstances of life were for the dinosaurs that lived there. On the upper part of the plain, nearer the mountains, the land was dry. To the east, the sea was probably 200 miles away. To the west, I would guess the mountains were 60 miles away. There was little rain; the land was semiarid. There was no grass; it hadn't yet evolved. But, from the remains of pollen left in rock, we know there were flowering trees (dogwoods), evergreens, berry bushes and huge, palmlike plants called cycads. There were numerous small streams with heavy vegetation on their banks—mostly dogwoods and evergreen trees, I suspect. Today, in this kind of landscape, we would expect grass to fill the large, flat expanses between the streams. Then, there were fruited plains of berry bushes. In spots (and at certain time periods during the 12-million-year span) there were shallow lakes that dried up each year to leave a dry, crusted surface, or hardpan.

If, during the time when the plain extended a full 200 miles, you were to have walked east from the mountains, you would have noticed a very gradual change in the land. As you crossed the plain, you would have felt yourself to be on flat ground—perhaps not quite as flat as Indiana or Kansas, but as flat as it is today in Montana near the Rockies. Slowly the narrow streams would have widened and joined with other streams to form meandering rivers bordered by swamps. The land would have become greener and dotted with ponds. The vegetation and the dinosaurs would have become more various. And after you had walked about 200 miles, you would have reached the sea itself.

It is the upper part of this plain, the dry land and small streams, the dogwoods, evergreens and berry bushes, that is preserved in the Two Medicine formation. Several kinds of dinosaurs lived on that plain, in that vegetation. The predominant ones were duck-billed dinosaurs, feeding on the evergreens and bushes. There were also small, graceful dinosaurs—the hypsilophodontids. And there were the ceratopsian or horned dinosaurs, as well as large carnivores that looked something like *Tyrannosaurus rex*.

The lower, greener, swampier part of the plain, from exactly the same time and also preserved in rock and fossils, is called the Judith River formation. Similar sorts of dinosaurs would have lived here—duckbills, ceratopsians, carnivores—but the more varied and abundant plant life would have meant a greater diversity of animal life as well. For the dinosaurs, that meant more species and much larger populations. The reason this lower plain is called by a different name and considered separate from the Two Medicine formation is that between the two formations is a dome, a bulge in the skin of the earth. Like a bubble in a viscous liquid coming to a boil, this bulge has been rising up intermittently since before the Cretaceous. It probably bulged most vigorously during times when the mountains were thrusting up. The result is that the middle of the plain we've been talking about is not preserved. Instead we find this bulge, known as the Sweetgrass Arch. The deposits that built up the coastal plain may have drifted off the dome as soon as they landed on it, or they may have been shed during a later thrusting up of the dome. They may have been washed away, or blown away, or scraped off by the glaciers—the bulldozers of the Pleistocene—that scoured all of Montana. Whatever happened to them, these deposits are gone, and what sits on top of the dome is the old bottom of the Colorado Sea, complete with fossil clams and snails.

Today you can find the eastern end of the Two Medicine formation and the beginning of the Sweetgrass Arch in the town of Choteau. The arch extends east about 60 miles, all of it flat. At the end of that stretch you come to the remnant of the coastal plain—this time the lower coastal plain—known as the Judith River formation. And again, instead of fossils of marine shellfish, there are dinosaur bones.

IN HINDSIGHT THERE WERE numerous clues to suggest that the Two Medicine formation would have been a good place to look for baby dinosaurs—in fact, a much better place than the Bear Paw shale. Certainly, everybody knew that early paleontologists had taken a lot of dinosaurs out of the Two Medicine formation. And there had been reports that some of these dinosaurs, originally identified as adults, were actually juveniles. In 1976, for example, Peter Dodson of the University of Pennsylvania published a paper on fossils found by several paleontologists in the Two Medicine and Judith River formations and previously identified as adults; he had studied the fossils himself and had determined that they were really juveniles of a completely different species.[7]

That was one clue. Another came from C. M. Sternberg, one of the great dinosaur field scientists, who had written a paper on fragments of young dinosaurs recovered from a formation in Alberta. This formation preserved an upper coastal plain of the same age as the Two Medicine formation. Sternberg argued that just such uplands might be the place to look for young dinosaurs. Referring to the fragments of young he had found as well as the rich deposits of young from Mongolia, he wrote:

> Many fine dinosaurs and other vertebrate fossils have been collected from the delta deposits of the [Judith River formation] along Red Deer River in Alberta, but very few fossils of juvenile dinosaurs have been reported. With the great number of experienced collectors who have examined these beds, surely some eggs and many

> remains of juveniles would have been reported if the dinosaurs had hatched and spent their whole lives on the deltas or in the swamps.

After describing the lowland deposits, Sternberg pointed out:

> ...no skeleton or skull of a very young dinosaur has been reported from these beds. This coupled with the fact that the upland deposits of Mongolia yielded so many dinosaur eggs and juveniles, leads one to believe that the dinosaurs laid their eggs on the upland and only the more or less mature animals of certain forms inhabited the deltas and flood plains.[8]

Both of these are, of course, isolated bits of information that there would have been no reason for me to connect, or even notice, back in the winter of 1978 when I was cataloging fossils from marine sediments. There was, however, another clue that surfaced just the summer before Bob Makela and I stumbled on the fossils in Marion Brandvold's rock shop. That piece of evidence was an intact dinosaur egg, the first found in the Western Hemisphere. I found it myself. And I found it in the Two Medicine formation.

My father and I had gone exploring for fossils. Our trip was something of a replay of one we had taken when I was a child; it was in the same formation that I had found my first dinosaur fossil, with my father, when I was seven years old. In 1977 we were again walking the ridges, looking for bones. I picked up what appeared to be a crushed lump of fossil bone and took it with me. I had no idea then what it was. Over that winter, however, I realized that it was a dinosaur egg, although I didn't know what kind of dinosaur it had come from.

Now, it might seem that if one found the first dinosaur egg in the Western Hemisphere, one would certainly go back and look for more.

And perhaps I should have. But one of the odd things about paleontology is that you can find *one* of anything almost anywhere. On that field trip when I found the egg, the Two Medicine formation appeared to me as it always had before—bare. Fossils were very, very hard to find, and there were no indications of substantial deposits. I had found one egg, but I might well have gone back and spent summer after summer walking over it and never have found another. What paleontologists look for is a pattern—not single fossils, but hints of widespread fossil deposits. To me the egg was an anomaly, whereas I had found a clear pattern in the fossils from marine sediments. The Bear Paw shale, obviously rich in fossils of juvenile dinosaurs, seemed to me the place to go. After all, I had not come up with the idea of looking for fossils of juvenile dinosaurs and then set about to see where I might find them. I had stumbled on a predominance of such fossils in the Bear Paw shale, and it was that predominance that gave me the idea of looking for babies in the first place.

Only much later did I see the pattern that tipped off the presence of these rare fossils in the Two Medicine formation. Only much later did I realize how likely a spot for babies the Two Medicine formation was. During the winter of 1979, after the trip to Marion Brandvold's rock shop and our follow-up of that trip, I checked all the fossils from the Two Medicine formation in the collections of Princeton University, the American Museum of Natural History and the Smithsonian Institution. Then I found out what I hadn't realized before—and what I think nobody else had realized, either. Eighty percent of those fossils were of juvenile dinosaurs, many of them unrecognized for what they were. When the fossils had been found originally, many had been identified as adults of new genera and species, or not identified at all, and shipped back to museum cellars. For example, in the 1930s the paleontologist Charles Gilmore found seven or eight small duckbills in one pit in the Two Medicine formation. He abandoned the hole, even though he thought he might find more, because the skeletons were all the same size and of the same kind of dinosaur.[9] Gilmore treated the skeletons like trout for the frying pan. He had enough, so he quit. It did not occur to him that he might have found some kind of social grouping of young; he didn't think they were young at all. He reported them as small adult dinosaurs of the genus *Procheneosaurus*, which had been described a few years before. Not so. They were young dinosaurs, though of what genus I'm not sure.

One can indulge in hindsight forever. There is no end of previously undiscovered clues and patterns. But the simple fact is that those first duckbill babies were found, not by painstaking analysis on my part, but by sheer luck. Marion Brandvold was lucky to find them. Bob and I were lucky to find her. And we were perfectly happy to give up our plan to go back to the Bear Paw shale. In paleontology, as in all sciences, as in all of life, you don't argue with luck. Marion Brandvold had discovered a lovely little window on the late Cretaceous. What we did was to open that window and climb through it.

CHAPTER 2

THE FIRST NEST

With a coffee can's worth of baby dinosaurs in hand, I called Don Baird at Princeton. I was on my vacation, after all, and I was due back at the university in a few days. Don extended my vacation and wired me $500 for expenses. Suddenly, instead of a preparator on a holiday, I was the principal investigator on a funded paleontological expedition—a small expedition, two guys in a van, but an expedition nonetheless.

The first thing Bob and I did was to go with the Brandvolds to the spot where Marion Brandvold had found the babies. The location was a cattle ranch near Choteau, at the eastern boundary of the Two Medicine formation. The bones had come from a little knob of mudstone no more than 4 feet high and 10 feet in diameter. To anyone not used to hunting for fossils, the top of that bump in the landscape would have looked like nothing at all. The whole knob was covered with gnarly limestone pebbles, fragments of mudstone, shards of other sorts of rock, and a few odd-shaped bits of gray-black stone.

Those bits of stone, a half-inch to a few inches long, were fossils. Some were lying right on the surface of the knob; others were partly in the mudstone and partly exposed. Bob and I got down on our hands and knees and gathered everything we could from the surface. When we sorted out the fossil bones from the dirt and the rocks, we found that we had the remains of two more baby dinosaurs.

The obvious next step was to see what was *in* the ground. We went back to the Brandvolds to see if we could get permission from the landowners to dig on the land. We were being very cautious. All we planned to do—at first, anyway—was to take out shovels and picks and rock hammers and ice picks to dig down into the mudstone and separate out the bones. Paleontology, however, is not only a matter of digging; it's also a matter of diplomacy and politics involving dealings with amateur collectors and landowners. And in the West, there's no politics like the politics of land. This land was owned by the Peebles family. The Brandvolds already had the family's permission for their own prospecting, and it seemed appropriate to go through them.

On August 9, two and a half weeks after our first stop at the rock shop in Bynum, we took some shovels and burlap bags and went out to put a small hole in the Peebles' pasture land. We worked very quickly because we didn't know how long we'd be able to stay on the land. We hadn't talked to the Peebles ourselves, and it wasn't clear from what the Brandvolds told us how welcome we were. So, instead of slowly and carefully uncovering each bit of bone, we just dug out big chunks and put them in burlap bags.

Our haste had one serendipitous result. By digging straight down, we exposed a clear demarcation line between green mudstone, with fossils in it, and surrounding red mudstone. Once we had dug out the green mudstone, the hole that was left had the shape of a hemisphere. It was a concave depression about six feet in diameter and about three feet deep, something like a giant salad bowl. The bowl itself was red mudstone, but the salad (which contained the fossils) was the green mudstone that had filled the bowl. When we realized this, it occurred to us that what we had was not just a collection of bones but a nest. It looked as if a mother dinosaur had scooped out a hole to lay her eggs in, a hole that was later filled in by a deposit of different-color silt.

We worked there for three days, digging up what we could and filling the burlap bags. Then we carried the bags back to our first field laboratory, Bob's backyard. Our tools were window screens and a garden hose. Paleontological laboratories are not as grandly equipped as organic chemistry laboratories or those of molecular biology. Nonetheless there are usually plenty of sinks and big tables where you can spread out the fossils you're working on. You have dissecting microscopes, and a variety of dental tools (paleontologists borrow from all the other disciplines) for scraping away dirt and stone from the fossilized bone. Tubs and trays are there for mixing the plaster to make

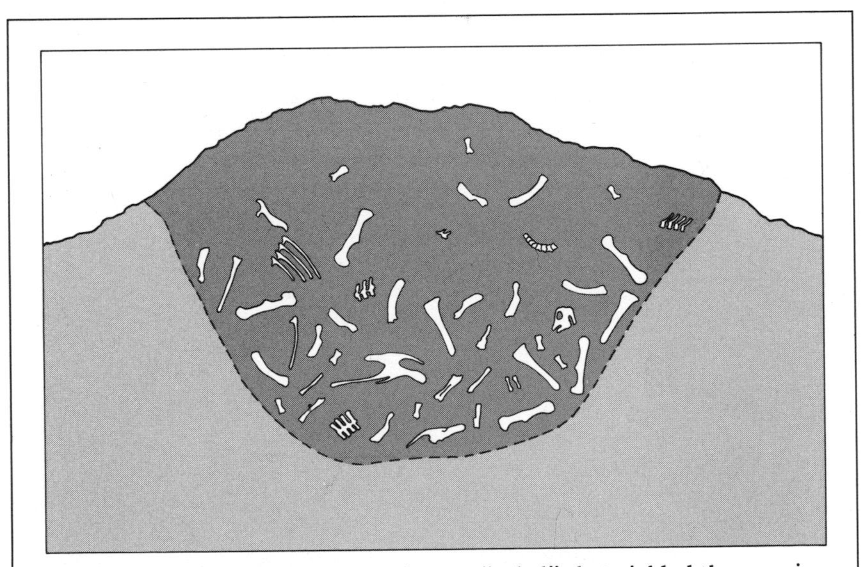

A cross section of the green mudstone "salad" that yielded the remains of 15 baby dinosaurs.

casts or for soaking fossils in acid baths to get rid of limestone. In short, a laboratory is a lot easier to work in than somebody's backyard.

To make the working conditions even stranger, we were being filmed by local television crews. The news media had found us even before we put shovel to earth on the Peebles ranch. Bob and I had been sitting in a bar in Shelby. We had already done our collecting from the surface of the knob and were speculating on what we might find once we started to dig. Somebody in the bar overheard us and telephoned Radio KSEN in Shelby to get a cash reward for a hot news tip. Whoever it was told the station that we had found a new dinosaur. This shows you how things work in Montana: not only is finding a dinosaur hot news, but Shelby is small enough that whoever telephoned the station was also able to tell them who Bob or I was and they were able to track us down at Bob's house in Rudyard. We convinced them to wait until we'd actually dug something up, but by the time we had the fossils and the dirt back at Bob's we had television interest, too.

This was the very first time I had to deal with reporters. After a while I got used to it. Newspapers and magazines started to pick up the story after I went back to Princeton, and more and more reporters came out as the size of the dig on the Peebles ranch grew from year to year. Television crews became, if not commonplace, certainly not exotic. By the time the dig had been going five years or so, we were getting two or three hundred visitors a summer, some from the press and some just curious. I finally had to hire somebody to do public relations, to take care of the merely curious, while I dealt with the press. But that first summer the attention was completely new and slightly bizarre. There we were, in Bob's backyard, with garden hoses and window screens, washing and starting to prepare our fossil find, with television cameras filming us as if we were doing neurosurgery on the president of the United States.

What we were doing was separating fossils from dirt and mudstone. First we screened the small bones from the dirt. When we finished, we had the remnants of 15 three-foot-long baby dinosaurs. To a nonpaleontologist, these remnants would have looked like nothing more than a bunch of black, stick-like rocks—jumbled and inscrutable, the way much of modern art seems to me. (Of course, I never studied art and I did study dinosaurs.) To me the bones spoke volumes.

I should say at the outset that one question I still don't know the answer to is *how* the babies died. Perhaps they died of illness. Perhaps their mother had been feeding them (an idea I'll get to later) and she died, leaving them to starve to death. I just don't know, and there is nothing in what we found to answer this question. I can, however, be more informative about what occurred after the babies died.

The fossil skeletons that we found were not articulated, which means that the bones had come apart at some point before the fossilization process began. Otherwise, we would have found neatly arranged skeletons instead of jumbles of bones. Furthermore, there were no marks or breaks showing that the bones had been cracked or chewed by predators prior to fossilization. These bits of evidence led me to think that after their deaths the babies must have lain rotting in the nest, slowly being denuded of flesh, and then, in a second stage of decay, their skeletons must have come apart as the connecting tendons and ligaments disintegrated. Eventually the bones would have also disintegrated from decay and the action of wind and rain if they had not been buried. But the nest, as was obvious from the kind of mudstone it was embedded in, had been in the floodplain of a stream. It had been buried by the muck of a spring flood when a nearby stream spilled over its banks.

Then, at some point after burial, fossilization began. The bones changed. Ground water seeped in and carried the mineral silica with it, filling in cavities in the bones. In addition, a process called replacement occurred. (This doesn't happen with all fossils, but it did with these.) Calcium and phosphates, the original molecules that the bones were made of, were leached out and the silica took their place. The bones became rocks. But although the materials changed, the essential structure remained. A fossilized bone that has undergone replacement—or mineralization, as it is also called—is like a baseball team that has had a lot of trades. The players change, but the positions remain. When you look at these fossils of baby dinosaur bones under a microscope, you can still see the canals for blood vessels and the minuscule structure of the bone itself, evidence of how it grew and what the physiology of the animal was like. The rock of the fossil is shaped, down to the tiniest detail, like the original bone.

The first and foremost tool of paleontology is comparison. The discipline is partly a collective memory of what this femur and that femur look like. I knew these were dinosaurs because femurs are distinctive for dinosaurs. I knew they were duckbills because I had pored over the collections of duckbill bones in museums and knew what duckbill femurs and tibias look like. And I knew they were babies because the bones were so small and because of their stage of development. Bones in dinosaurs, indeed in all creatures, follow a predictable pattern of growth, and these fossils showed specific signs of extreme youth. For instance, the vertebrae near the base of the spine, which in the adult are fused to form a solid section called the sacrum, were not yet fused. A second clue also concerned the

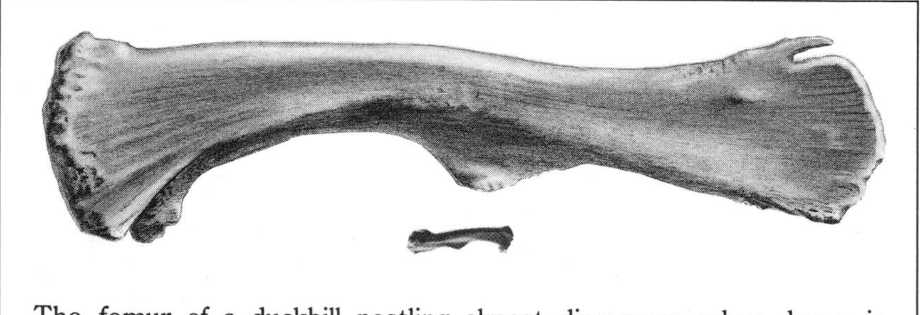

The femur of a duckbill nestling almost disappears when drawn in relation to the adult femur.

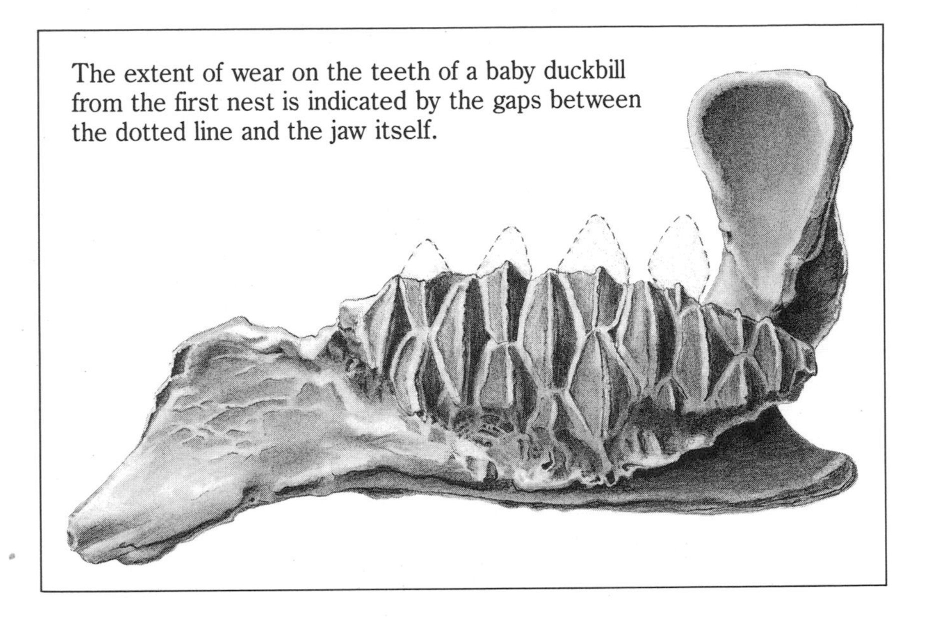

The extent of wear on the teeth of a baby duckbill from the first nest is indicated by the gaps between the dotted line and the jaw itself.

vertebrae. Each vertebra is composed of two parts: the centrum and a smaller bone, called the neural arch, that looks like a fin. In the adult these two parts are fused, but not in the young, and in the vertebrae we found in the mudstone the parts were not fused. Still another clue had to do with the limb bones. The ends of the bones were not yet fully formed, a sure sign of a very young animal.

The dinosaurs from that nest were not so young, however, that they had just hatched out of the egg and keeled over dead. The tendons that run along the spine and keep the tail off the ground were already hard, or ossified, not flexible as they would be at the time of hatching. And, an even stronger sign, connected not to development but to the environment, was that the teeth were worn—some of them almost three quarters gone. (Like sharks and most reptiles, dinosaurs wore out their teeth and had new ones come in throughout their lifetimes. In a well-preserved fossil jaw you can see the ranks of developing teeth waiting under the ones in use, like fresh troops ready to relieve weary combat veterans.) Clearly, the young had been eating for some time.[1]

Another way of interpreting fossils is to use the indirect evidence of the environment, the geological evidence on both the large and the small scale. Often concentrations of fossil bone are deceptive; one can be fooled into thinking that the animals found together in death were also together in life, when, for example, what really happened was that they all died near the same stream at different times and were carried by the flow of water to the same spot. In the case of these babies, however, I don't think we were fooled. Remember, they were found in a bowl of green mudstone that was set off by a sharply defined boundary from the surrounding red mudstone.

We later found this same form repeatedly with heavy concentrations of fossilized dinosaur eggshell and, sometimes, fossilized bones of baby dinosaurs in the green mudstone. The only green mudstone in that area was in these bowl-shaped hollows containing either eggshell or bones and eggshell together. The hollows might have been accidents, but the repetition of the shape, always with bones or eggshell and always with green mudstone, made the likelihood of that possibility almost zero. I'm sure the hollows were nests. As I said before, I think first the mother dinosaur dug a place for her eggs in the red mud; then much later, after the eggs in the successful nests had hatched and those babies were gone, and after the babies we found had died, stream flooding filled up the hollows with a different sediment—the green mudstone. Another fact in favor of this explanation is that all the dinosaurs in that first nest were the same size. This means they were the same age, which, if they hadn't hatched together in the same nest, would be a very unlikely coincidence.

If all this was true, the next step in interpreting the fossils was to explain how they could have hatched together, stayed together, eating all the while, and then died in the nest. The most likely explanation is that they had never left the nest, and that one or both parents cared for them, bringing them food. It's conceivable that they wandered around in a group, feeding outside the nest and then returning there to rest, but I find this hard to believe. For one thing, this would have meant 3-foot dinosaurs walking among their 30-foot parents. That's a dangerous way to start life. I think they would have been safer in the nest.

And yet if they were coldblooded, like modern reptiles, it would have taken them a long time to grow. They would have to have been in the nest for a number of months, perhaps almost a year. This is not something anybody has seen happen in any kind of existing reptile or bird. Living reptiles just don't do their growing in the nest; they get out as soon as they hatch. And warmblooded creatures such as birds, which do grow quite a bit while in the nest, grow faster. The inference we drew from these facts was that these animals were doing their growing in the nest, that they were probably doing it fast—and that they were therefore probably not coldblooded like most modern reptiles but warmblooded like birds and mammals.

ALL THESE CONCLUSIONS about the nest, about the behavior of these young dinosaurs, about their physiology, came from the bones of the babies and from reading signs in the rock they were found in. But none of these things told us what genus or species of dinosaur we were dealing with. They were duckbills, to be sure, but what kind? To identify dinosaurs one usually needs a very distinctive bone or set of bones, and the most distinctive set of bones you can find is the skull. I haven't mentioned it until now, but there was another fossil we brought back to Bob's backyard: a skull, and a telling one.

The Brandvold family had found it. While we had been digging up the nest of baby bones, they had been attacking another nearby fossil deposit with pick and shovel. By the Brandvold family I mean Marion and her husband John, and David and Laurie Trexler. David, Marion's

Baby maiasaurs, each equipped with a small, sharp "egg tooth" on the end of

its nose, crack their shells to emerge in the security of the nest.

son by a previous marriage, had married Laurie, who was John Brandvold's daughter by a previous marriage. All of them prospected occasionally for fossils and all were working on this particular deposit. On August 11, after we'd finished our own work, the Brandvolds asked us to look at something. It was a skull, still in the ground and pretty badly bashed up. They had been trying to get it out of the ground in one piece, but they hadn't been successful. Although some parts of the skull were already in fragments, we managed to salvage it. We cleaned the visible parts that had not been broken up and found that there was, at the least, a complete snout with a telltale duckbill. We painted the skull with shellac to hold it together and then covered the whole thing with burlap and plaster of Paris to make a jacket, or cast. The next day, when the jacket had hardened, we went back, got the skull out of the ground and brought it, securely jacketed, to Bob's backyard, where we worked on it along with the baby bones.

We gently removed the plaster and washed away the dirt. We didn't know for sure that the skull and the babies belonged to the same genus or species because, as I said, we had no baby skulls. That came later. We did know, however, that they had been found together in the same deposit. And we could see, clearly, from the material we had that they were both duck-billed dinosaurs. We knew that some adults had to be around caring for the young dinosaurs. It seemed highly likely that the babies and the skull were from animals of the same species. This adult may not have been a father of the babies, but my guess was that it was some kind of uncle.

The shape of the snout and the position of the nose hole, which I'll describe in detail in the next chapter, made it quite clear that this skull represented a new species of dinosaur. In itself, finding a new species of dinosaur is a nice discovery but not all that uncommon. A relatively undramatic difference in the shape of a snout may identify a new species, but the find is not that impressive if this species lived and acted like similar ones. What makes a find more or less important is what there is that's new and different about the dinosaur. Is it just that it has longer legs than other dinosaurs? Or, for example, is there something about it that suggests some new and exciting behavior? Was it a particularly fast runner? Or did it have a weapon-like hook on its nose that it used to slay prey? (This is a fantasy; there's no dinosaur like that.) In this case, we were fairly sure we had found a creature with an extraordinarily interesting new behavior, completely unknown for dinosaurs: parenting.

This was the first time anyone had found a nest not of eggs but of baby dinosaurs, and the evidence seemed to me incontrovertible that these babies had to have stayed in that nest while they were growing and that one or more parents had to care for them. This kind of behavior, unheard of in dinosaurs, was probably the most startling discovery to come out of that dig. Certainly it was the one that had the greatest effect on the public image of dinosaurs, because it was in such severe contrast to the image of how dinosaurs were supposed to behave—laying eggs and leaving them, like turtles or lizards or most reptiles. If dinosaurs, even just some species of dinosaur, had acted like birds and reared their young in nests, caring for them and bringing them food, this was a bit of information that would profoundly change our sense of what sort of creatures these ancient reptiles were.

It was a revelation, and it was only right that this behavior should be the source of the new dinosaur's name. Don Baird invented the appropriate name for the dinosaur, and Bob and I christened it formally when we described it in print. We called the creature *Maiasaura peeblesorum*. The species name, *peeblesorum,* comes from the owners of the land on which the fossils were found, the Peebles family. The name that tells the story is the first one. *Maiasaura* comes from the Greek and means, roughly, "Good mother lizard."

Chapter 3

The Good Mother Lizard

If we are to understand who this good mother lizard was, what she looked like, where she came from, the first thing to be said is that she was a duck-billed dinosaur. (It seems only right, in view of the meaning of *Maiasaura,* to use the feminine pronoun.) But then that raises more questions than it answers. Who were the duckbills? Where did they come from? Where does *Maiasaura* fit among the duckbills? Where do the duckbills fit among the dinosaurs?

Perhaps it's best to start with the dinosaurs themselves. The history of the dinosaurs begins a little over 200 million years ago, in the Triassic period, when the planet was one large landmass set in one giant ocean and the land was dominated by large, flesh-eating reptiles.[1] Near the seas and rivers the land would have been green and fertile, heavily forested with conifers and cycads, which looked something like palm trees. But in most areas the climate was dry and vegetation would have been less lush, consisting of evergreen bushes, sedges, or trees that could tolerate conditions of low rainfall.

This was the world in which the dinosaurs evolved. In their turn, they became the dominant large land animals. And they retained this dominance for 140 million years, until the mass extinction at the end of the Cretaceous period. The first appearance of the dinosaurs was not, however, as dramatic as their later domination of the planet. Their evolutionary debut was marked by nothing more than the slight turn of a hip.

The traditional language of evolution uses a kind of shorthand. Fish evolved into amphibians and amphibians into reptiles and one group of reptiles into another group of reptiles called dinosaurs. From this or that kind of dinosaur, these others emerged. This is all correct, but without some background knowledge the process sounds like the transformations of a magician—the handkerchief turns into a bouquet of roses and the roses into a dove. The reality is somewhat different.

Changes in genes, in the DNA that carries the instructions for growing any given organism, provide a variety of sizes, shapes and muscles, of innate capacities for speed or acuity of eyesight among organisms in the same species. Some Thoroughbreds run like the wind; some are stumblebums. The better a male horse runs, the higher the stud fee his owners demand when he moves on from racing to reproducing. The fee is for his genes. Of course, with racehorses we select which individual animals get to pass on their genes; we test them at the track, and the fittest, by our standards, get to spend their lives as broodmares and studs. That's artificial selection. Natural selection is somewhat similar—Darwin called it survival of the fittest. But in natural selection fitness is defined not by speed or intelligence, although these may be significant, but by how many viable offspring an organism leaves. If you happen to be a garden slug, for instance, and your coloring makes you less susceptible to murder by the gardener, you may live longer and your offspring may be more numerous. It's no accident that garden slugs are not fire engine red.

These days there is some discussion about the importance of

natural selection, whether some evolutionary change is random in nature, and not selected at all, and what the pace of evolution is. Are there rapid bursts alternating with long periods of little change? Or is the progress a steady and slow one? Still, the broad outline of evolutionary change is clear. And a couple of points are important to note before I discuss how the dinosaurs evolved. The transformation of roses into dove is not a good model for evolutionary change for two main reasons. The transformation is too quick, and the roses simply disappear.

The amphibians did not just disappear when the reptiles emerged. Some, to be sure, evolved into reptiles. That is to say, certain lineages of amphibians experienced certain evolutionary changes in skin and breathing apparatus and skeletal construction. At some point, when the resulting creatures were different enough to pursue life on land without needing ever to return to the water, they obviously required a different classification. In contrast to amphibians, reptiles need not live in watery environments (although some do, just as some mammals do).

We could say that the lineages of amphibians that evolved into reptiles disappeared, but really they continued on in different form. Other lineages faced with competition from the new reptiles simply went extinct, with no descendants. This is true disappearance. Still others stayed the distance, evolving and changing in some ways but remaining amphibians. From frogs to salamanders, the world is full of amphibians today—not the big, fierce, carnivorous amphibians that once prowled or perhaps sprawled through the conifers in the time before the heyday of the reptiles, but amphibians nonetheless. This is true for many forms of life. Certainly early single-celled organisms gave rise to all life on earth. But we've still got plenty of single-celled organisms like amoebas that have themselves been slowly evolving to become better adapted to their environment. In evolution the fates of ancestors are various.

TO SKETCH THE RISE of the dinosaurs, we can begin with their immediate ancestors. These were the reptiles called thecodonts, most of which were meat-eaters. Some species of thecodonts had developed a new, rapid means of locomotion. They walked and ran on two legs, instead of sprawling along on four like overgrown lizards as had all reptiles and amphibians before them. The dinosaurs went on to improve this form of locomotion. The first dinosaurs, carnivores like their ancestors, were rapid, efficient, two-legged runners. Later, some forms of the dinosaurs developed four-legged or quadrupedal locomotion, though in a more stable and efficient form than other reptiles had achieved. The clearest difference between the first dinosaurs and the last thecodonts lay in the hip socket. Dinosaur sockets are open (there is a visible hole in the bone into which the thighbone fits), and thecodont sockets are closed.

From the time the first dinosaurs emerged (*Staurikosaurus* is the earliest known), their history was one of expansion and diversification. Their 140 million years on earth spanned three geologic periods. After the Triassic period came the Jurassic, when some of the favorite dinosaurs of children evolved—the sauropods, the biggest dinosaurs of all, in fact the biggest land animals ever to exist. The most famous of these, *Brontosaurus*, is known to almost everybody. (It is, however, known by the wrong name. *Brontosaurus* is now correctly called *Apatosaurus*.) Other sauropods had a similar look: thick, pillar-like legs, long, almost snakelike necks and equally long tails. The stegosaurs also flourished during this time, as did *Allosaurus* and other large carnivores that were the precursors in form of *Tyrannosaurus rex*.

Then came the Cretaceous, which is probably when the dinosaurs reached their greatest diversity. The horned dinosaurs, including the familiar *Triceratops*, were widespread. Club-tailed dinosaurs called ankylosaurs also thrived. A wide variety of smaller dinosaurs and predatory dinosaurs of every size flourished in North America, Asia, just about all over the world. And it was during the Cretaceous that the dinosaurs of most interest to us, the duckbills, appeared. They themselves then diversified.

The dinosaurs, along with all the other reptiles that are living or have ever lived, are all part of the class Reptilia. Within this class the dinosaurs occupy different orders, genera and species. When fossils of dinosaurs were first found, in the early 1820s, they were something of an anomaly. Scientists were confronted with fossils of three large, terrestrial and (two of them) herbivorous reptiles that did not quite fit with any of the known reptiles, living or extinct. For one thing, no large herbivorous land reptiles were known. Also, all the new animals had in common an open socket where the femur joined the pelvis, and, finally, they all displayed a new look in the way the pelvis was joined to the spine. So in 1841 Richard Owen, the first head of the British Museum of Natural History, took a good hard look at *Iguanodon*, *Megalosaurus* and *Hylaeosaurus*, as the first three of the new reptiles had been named, and created a new class: Dinosauria. The name means, as every schoolchild learns, "terrible lizards."

Biological classification is, however, full of pitfalls, particularly when the evidence for creating a new group is small. It is also a matter of consensus among paleontologists. There is no absolute proof that this or that classification is correct. Scientists like Owen, who wish to change the system of classification, must amass their evidence, publish it, and try to convince other paleontologists. Owen succeeded, momentarily. But he was dealing with odd bones and fragments, not with skeletons or partial skeletons. In the late 1800s paleontologists determined that there were two separate orders of dinosaurs, the Saurischia and the Ornithischia. So the original class, the Dinosauria, fell out of favor.

The difference between the Saurischia and the Ornithischia has primarily to do with the pelvis. Saurischia literally means "lizard-hipped," and Ornithischia "bird-hipped," and you can see the difference in the drawing on page 71. As to evolutionary history, the saurischians

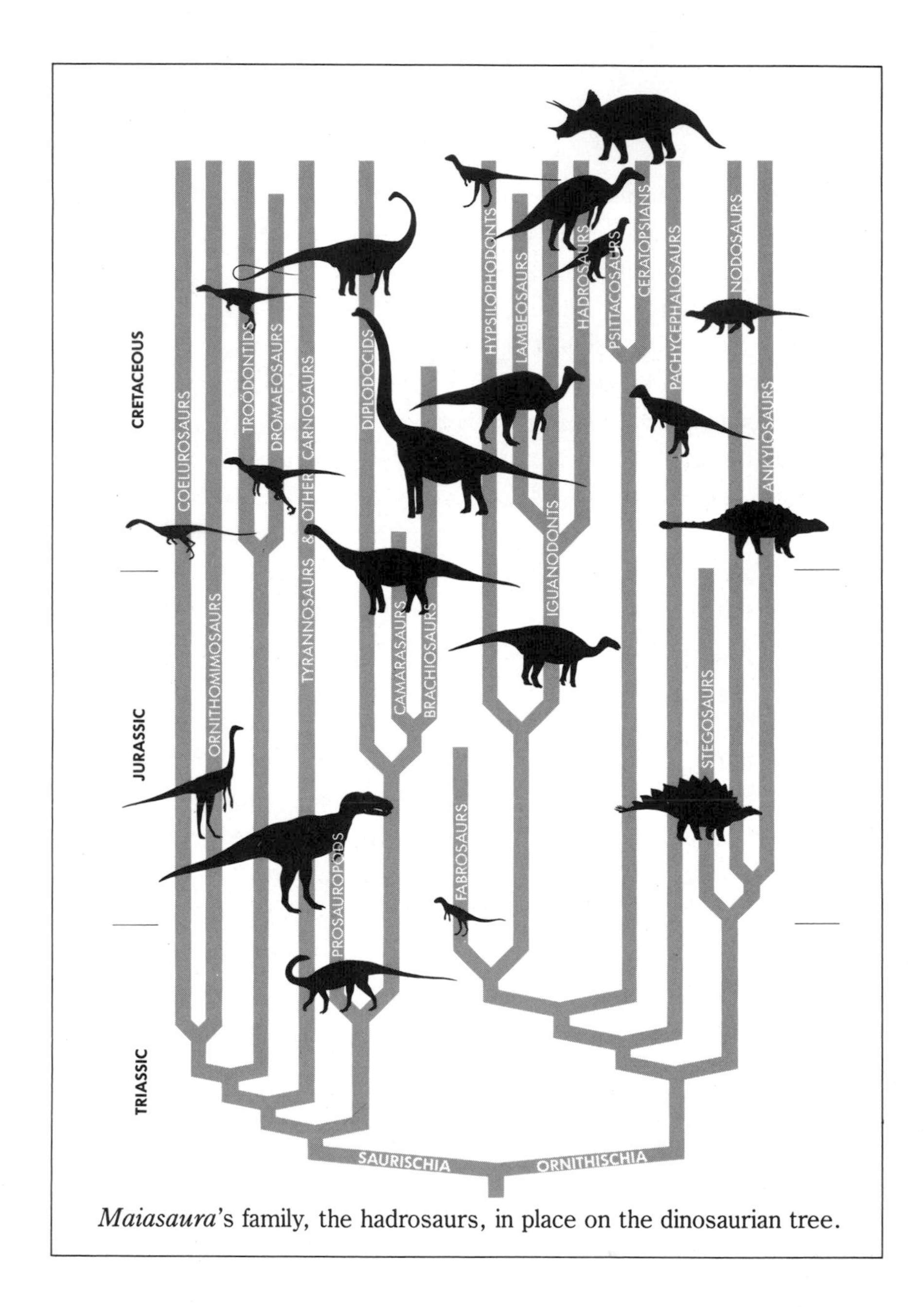

Maiasaura's family, the hadrosaurs, in place on the dinosaurian tree.

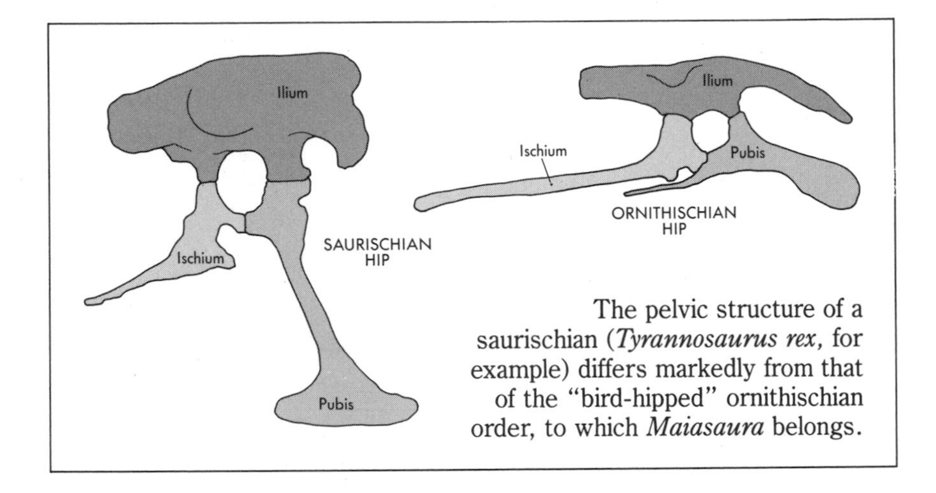

The pelvic structure of a saurischian (*Tyrannosaurus rex,* for example) differs markedly from that of the "bird-hipped" ornithischian order, to which *Maiasaura* belongs.

seem to have appeared first, in the late Triassic, about 200 million years ago. Shortly thereafter the first ornithischians appeared, perhaps descended from the saurischians or perhaps coming from some ancestral reptile that was similar to, but not the same as, the ancestor of the saurischians.

Within each of these orders, there are numerous different suborders, families and genera. To take some of the most common and well-known dinosaurs, all the carnivores—from the large and fearsome *Tyrannosaurus rex* to the small, quick and also fearsome *Deinonychus*—are saurischians. So are the sauropods, such as *Apatosaurus*. Furthermore, it now seems that birds descended directly from the saurischians. Partly for this reason, a variant of the original class Dinosauria has recently come back into favor. In the 1970s, Robert T. Bakker and Peter Galton published a paper arguing that Dinosauria was a legitimate class but that it should include not only both orders of dinosaurs (ornithischians and saurischians) but the birds as well (now alone in the class Aves). Their proposal has since been a lively subject of discussion in paleontology because what it means, in effect, is that there are still living dinosaurs among us—the birds.

As to the ornithischians, they include a number of suborders, among them the stegosaurs (with a ridge of triangular plates along the spine), the ankylosaurs (with armor and clubs on their tails), the ceratopsians (with horns, like *Triceratops*) and the ornithopods. Within the ornithopods are many families, including the one that interests us: the Hadrosauridae, also known as hadrosaurs or, in the vernacular term we've been using so far in this book, duck-billed dinosaurs. This is the family to which the genus *Maiasaura* (which includes only one species, *Maiasaura peeblesorum*) belongs. Duckbills, hadrosaurs and Hadrosauridae are all words for the same animals, a family of dinosaurs with what look like ducks' bills.

The duckbills appeared in the late Cretaceous. They were all herbivores and they all had a two-legged gait, although some may have used their forelegs to help them out occasionally—in rough terrain, for example. They also all had the snouts that resembled ducks' bills. By the time they evolved, the landmass had separated into continents, and the remains of duckbills are found in Europe, Asia, and North and South America. We don't know on which continent they first evolved. In whatever part of the globe they were, they lived on the coastal plains of one sea or another. (As did all the dinosaurs.) We don't know whether they, or any other dinosaurs, also lived in inland areas, because there are no geological formations that preserve inland habitats from the dinosaurs' time. Deposition or sedimentation, which is what makes sedimentary rock and what preserves fossils, did not occur to a large enough extent in the inlands. Consequently, the known habitat of the duckbills was more or less the same across the globe: the lush, green swamps of the low coastal plains or the semiarid upper coastal plains. The duckbills did not live alone, of course. The horned dinosaurs, the club-tailed dinosaurs and many other large predators inhabited the same territories.

In North America the duckbills first appeared around 100 million years ago during the recession of the Colorado Sea, the event that marks the beginning of the Two Medicine formation. Before that, the big bipedal herbivores were dinosaurs called iguanodontids—the ancestors of the duckbills. Once the duckbills appeared, they themselves evolved into a variety of forms. They and the ceratopsians were the dominant herbivores in the late Cretaceous—one walking on two legs, the other on four. They were preyed on by a variety of carnosaurs such as *Albertosaurus*, an animal that was a slightly smaller foreshadowing of *T. rex*.

Within the duckbill family, paleontologists have traditionally recognized two subfamilies: the flat-headed Hadrosaurinae and the elaborately crested Lambeosaurinae, known more informally as the hadrosaurines and the lambeosaurines.[2] The guess was that the lambeosaurines evolved from the simpler, less ornately decorated

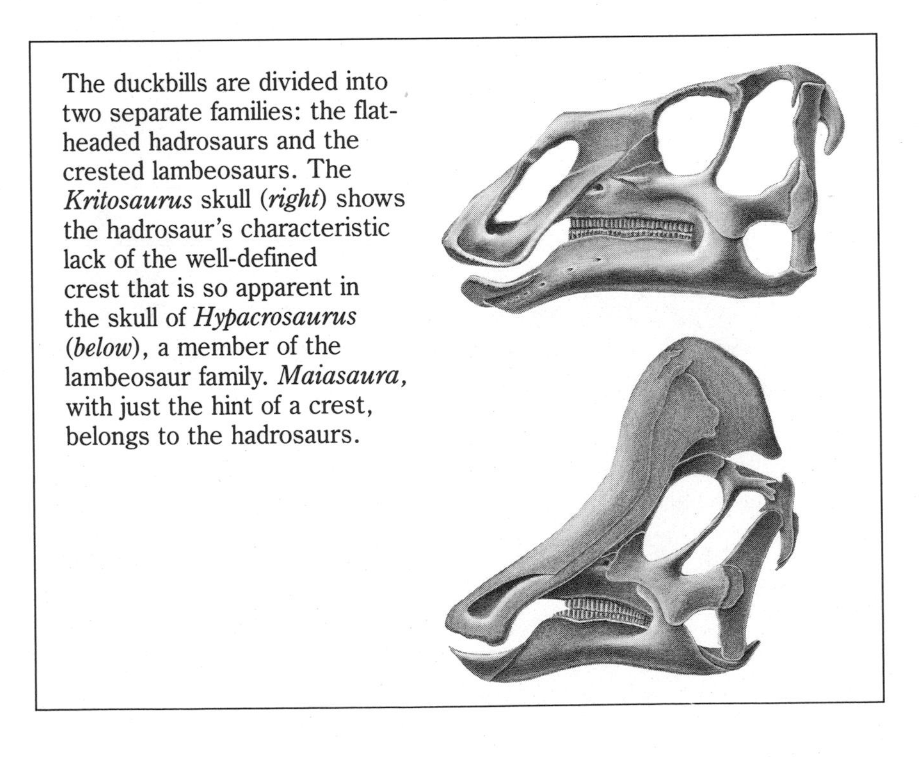

The duckbills are divided into two separate families: the flat-headed hadrosaurs and the crested lambeosaurs. The *Kritosaurus* skull (*right*) shows the hadrosaur's characteristic lack of the well-defined crest that is so apparent in the skull of *Hypacrosaurus* (*below*), a member of the lambeosaur family. *Maiasaura*, with just the hint of a crest, belongs to the hadrosaurs.

hadrosaurines. (*Maiasaura*, by the way, is of the plainer variety; she has no fancy crest.) However, I've uncovered some evidence suggesting that this classification is no longer valid.[3] The flat-heads and the crested duckbills each seem to have evolved from different sorts of iguanodontids, not from the same one. This is called polyphyletic origin, and it requires that they be considered separate families rather than subfamilies. All this means is that the scientific names for these creatures are now the Hadrosauridae and the Lambeosauridae, a change of one consonant. In colloquial terms we can call them the hadrosaurs and the lambeosaurs, and both are still duckbills.

THE FOSSILS OF DUCKBILLS have played a prominent role in the history of paleontology and in my personal history, in the circuitous route that I followed from my home in Montana to New Jersey and back to Montana to *Maiasaura*'s lair on the Peebles ranch. In retrospect, I can't imagine that the most important dinosaur fossils I've ever found could have come from any other kind of animal.

One of the very first dinosaur fossils discovered in North America was a duckbill tooth. It was found in Montana, in 1854, in the Judith River formation by paleontologist Ferdinand Hayden, whom the Indians thought to be insane and therefore holy. He found several different teeth, of different dinosaurs, which is why we can't say whether the hadrosaur tooth was actually the first. Hayden was out in the West during the Indian wars and sometimes found himself in the territory of Indians who, for good reason, were less than hospitable to whites. They had a name for him, and I think if you imagine a paleontologist caught unwittingly in some sort of skirmish and trying to get that last hadrosaur tooth pried out of the mudstone before he gets a bullet or an arrow in him, the name acquires its proper resonance: he was called "the man who picks up stones while running."[4]

The very first relatively complete dinosaur skeleton found anywhere was also of a hadrosaur. It was found in Haddonfield, New Jersey, in 1858 by William Parker Foulke, on a vacation from Philadelphia. Up until this time, most dinosaur fossils had been fragments and most, except for Hayden's teeth, had been found in Europe. Foulke found the skeleton in a marl pit and brought it to Joseph Leidy at the Philadelphia Academy of Natural Sciences. Leidy, the preeminent American paleontologist at that time, inspected, named and reconstructed the skeleton of the beast and called it *Hadrosaurus foulkii.*

After this, of course, hadrosaur discoveries burgeoned. Hadrosaur and lambeosaur bones are probably the most common dinosaur fossils found in the American West. And that's how I became involved, if not entangled, with duckbills. I literally followed in the footsteps of Hayden and other paleontologists who plied the North American Cretaceous deposits. And I found what they found—hadrosaur and lambeosaur bones.

There were no great fossil collections in the West when I was in school. Now there is the new $28 million Tyrrell Museum of Palaeontology in Drumheller, Alberta. In the size and scope of its vertebrate paleontology collection, it beggars every other museum in the world. But that's now. The museum wasn't built until 1986. In the late nineteenth and early twentieth centuries just about everything that came out of the ground, in the United States or in the Canadian West, was carried back east. So, although I had read the histories and haunted the same formations in which the early American paleontologists made their great finds, I had never seen the fossils they had collected. I had been to the Far East, to Vietnam by way of Camp Pendleton in California, but I had never been to the East Coast before I got the job at Princeton.

When I arrived in the East and went hopping from one museum to another with Don Baird, looking at the fossil collections, I found more hadrosaurs and lambeosaurs. For instance, Douglass' juveniles from the Bear Paw shale, which got me interested in looking for babies, were almost all duckbills. And there is one other quite important hadrosaur that I found in a museum.

At first, as a newcomer fresh from the territories, I was awed by the collections. But I soon found out that many of them had fallen into disrepair, that some of the great names I had read about had pulled stuff out of the ground without recording much information about it, so that it was now all but useless, and that in some cases the disorder and chaos of the collections were incomprehensible. One of the museums Don and I went to was the Philadelphia Academy of Natural Sciences. Today, it has one of the best and most modern presentations of dinosaurs in the world, but then it was a mess and hopelessly out of date. Nonetheless, I was entranced by it. It was the museum where some of the great early paleontologists had worked. Joseph Leidy, the first great American vertebrate paleontologist, had done his work there, and it was there that he reconstructed *Hadrosaurus foulkii*, the first dinosaur skeleton found.

I had already been to see the site where this partially preserved skeleton had been discovered. (It had become a housing development.) And now I wanted to see the reconstructed skeleton itself. But when Don and I asked to see *Hadrosaurus foulkii*, we were told it was lost. The skeleton wasn't gone. It hadn't been thrown out. It was lost *in* the collection, which gives you some idea of the collection's sorry condition. Two of the bones from *Hadrosaurus foulkii* were on display, set in concrete at the base of another dinosaur, but the rest had disappeared into the clutter. In the back rooms, away from the displays, there were bones lying on the floors and stuffed into little glass cabinets one on top of another in a jumble. There was no order that I could see.

I started going to Philadelphia regularly, sometimes with Don but usually by myself, to look for *Hadrosaurus foulkii*. And I began to find the bones. First I identified them by color—all the bones from New Jersey had the same black color. Then I went over the old records, tracing catalog numbers and trying out bones to fit them to each other.

Eventually I found the thing and put it back together. This was not exactly a field discovery; it was more a rediscovery. But putting that skeleton back together was enormously satisfying. I got the feeling that I myself was part of the history of paleontology, and it still gives me pleasure to see the reconstruction when I go to the Academy today. There are displays of *Maiasaura* nests there, and videos of me working in the field, an established paleontologist with a dig, a crew, and grants to support them. But there is also *Hadrosaurus foulkii*, restored by me, the preparator, who was glad to have any job at all in paleontology.

IT WAS ACCIDENT and availability that led me to the hadrosaurs and lambeosaurs, but there were numerous reasons to stick with them. The duckbills are among the most successful of dinosaurs. Numerous species of duckbills thrived during the Cretaceous all over the earth. In the late Cretaceous they were, with the ceratopsians, the dominant herbivores on land. Studies of the duckbills have long been a thriving paleontological business, so that working on them means entering a rich and varied field. And to my mind the hadrosaurs and lambeosaurs are two of the most sophisticated reptiles ever, living or extinct.

I say this because of their teeth. Teeth are very important in the study of fossils. They are hard, and therefore often well preserved, and they may have a lot to say about how the animal lived if you can decipher the clues they offer. The teeth of duckbills are dramatic. All species have some differences, but a typical duckbill jaw has scores of teeth, always being replaced, arranged in several rows on either side of the lower and upper jaw. These are not subtle clues. This kind of grinding apparatus was almost certainly used for processing plant food, just as the carnosaurs' curved, serrated teeth, like small (or sometimes large) steak knives, were used to cut and rend flesh. The

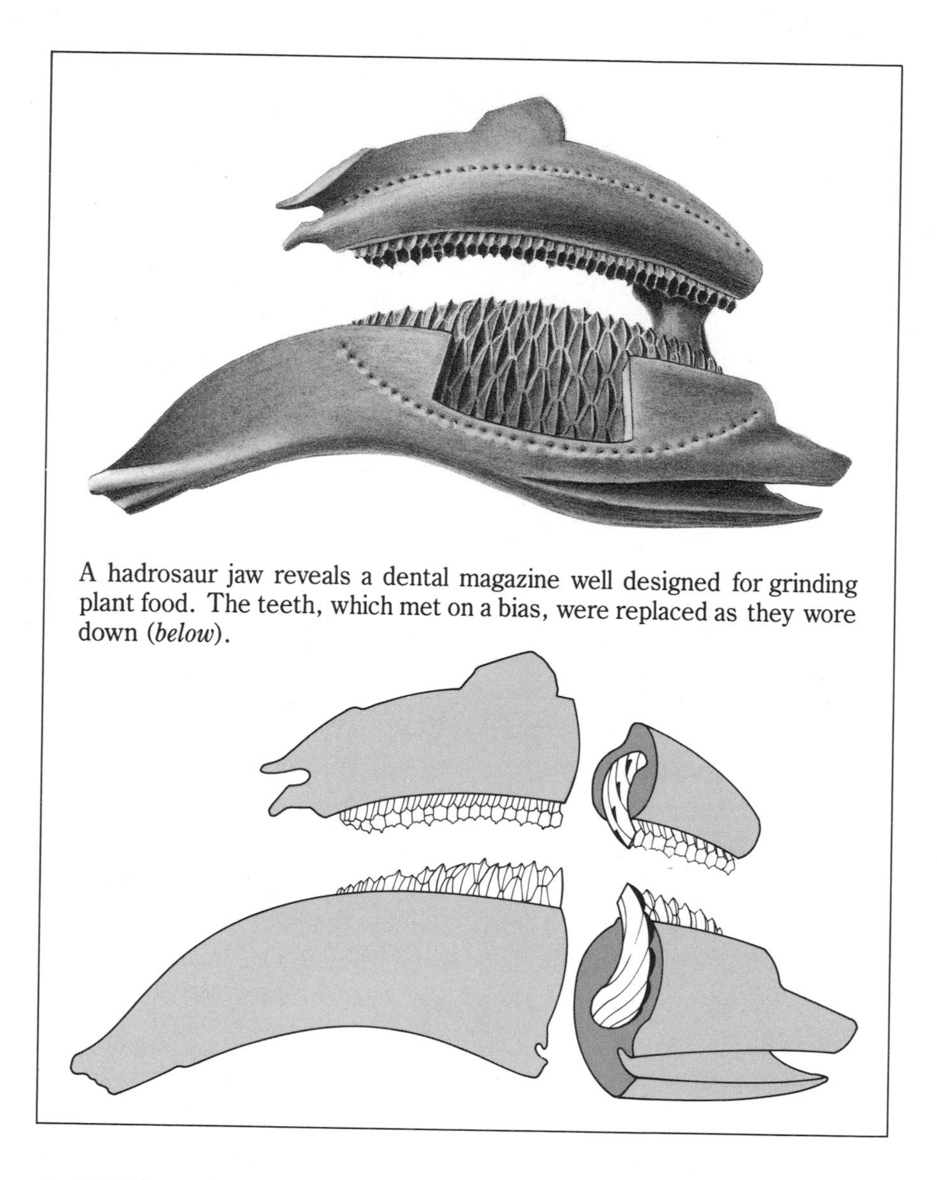

A hadrosaur jaw reveals a dental magazine well designed for grinding plant food. The teeth, which met on a bias, were replaced as they wore down (*below*).

duckbills' teeth were very well designed for herbivory. They were the rotary mowers of their day.

All this tooth talk may not sound impressive to a mammal. None of us has a dental magazine like the duckbill's, but we do have our share of impressive chewers. Goats and sheep do a respectable job, as do cows, buffalo, giraffes, horses, elephants—there's a long list. Reptiles have no such list. Among reptiles the duckbills are set apart not because they were great chewers, but because they chewed at all. Very few other reptiles, with the notable exception of the ceratopsians and iguanodontids, have ever been able to chew their food. Most reptiles can bite, cut, shear, chop and swallow, but not chew.[5]

An even more mundane quality that I find very appealing is that the duckbills have manageable skulls. If you want to look at a *Triceratops* skull, and I've done this several times, you practically have to dig a cave underneath it. It's seven feet long and weighs a thousand pounds. And the separate bones of the skull are fused together solidly. Duckbill skulls are small enough to pick up, and they come apart as well.

Originally, all the duckbills were thought to be aquatic dinosaurs, primarily because of their bills. Some of them also have webbed feet, and their broad tails look as if they would have been good for sculling. Then, in 1964, John Ostrom of Yale published a paper arguing that the tail was quite useful for balance, that the hadrosaurs' teeth were well designed for chewing tough terrestrial plants, and that hadrosaur fossils are as often as not found in terrestrial environments where there would not have been that much water.

I agree with him that many duckbills were probably purely terrestrial. But I'm also convinced that at least a few of them were semiaquatic. One big reason is the nature of their skulls. Some duckbills have kinetic skulls. This means that unlike human beings, who have movable lower jaws attached to rigid skulls, these dinosaurs had skulls in which *many* of the bones moved.[6] Birds also have kinetic skulls, all birds. The movable skulls come in particularly handy for ducks that feed on vegetation in the water. They take in plants and water, then press the upper jaw down on the lower one to push the water out through a strainer system and keep the plants in. One particular genus of hadrosaur, *Gryposaurus*, has a similar setup in its mouth. The edge of its bill is crenulated, or crinkled, making a kind of filter to trap plant material. *Gryposaurus* lived in a swampy area near the sea. It had webbed front feet, and it had a deep tail that would have worked well for sculling. Maybe *Gryposaurus* wasn't semiaquatic, but if it wasn't, no dinosaur was.

Whatever they did in the water, on land duckbills probably moved like birds, with their heads bobbing forward and back.[7] They did not look like the dinosaurs that have their tails dragging on the floor in the American Museum of Natural History and a lot of other museums. I've said this many times (and so have other paleontologists; it's not my original insight): the tails of most dinosaurs, not only the duckbills but also the sauropods, were held out straight behind them. The duckbills' tails were reinforced by rigid, ossified tendons that we can still see in many fossil skeletons. For the duckbills, the evidence is particularly good. There's a lambeosaur in the American Museum of Natural History with a neck curved almost like a swan's. It is displayed in a case as if it were swimming. But I think that curve in the neck, which is found in other duckbill skeletons preserved in articulated form, would have made sense if the animal had been walking. A swan's neck curves when it swims, but so does a goose's neck when it walks, or a pigeon's, or a chicken's. Watch the way birds seem to bob their heads forward and back when they walk; this redistributes the weight of the body, which is perched like a seesaw on two legs. Bipedal dinosaurs were built the same way. When a duckbill walked, I think it would have had a curved neck and a bobbing, fluid gait.

I don't know what duckbills looked like when they were running, perhaps like huge, leathery ostriches, but I'm absolutely certain they could run. That was one of their major defenses against predators. Other defenses might have been living in herds, a subject for another chapter, and, probably, the ability to deliver quite a solid kick with the hind legs.

I also suspect that the duckbills could make noise. There has been a fair amount of attention paid to the noses and crests of various duckbills. The crests of the lambeosaurs are apparently extended nares. In other words, the lambeosaurs added on, evolutionarily, to their nasal cavities and got, with the extra space, a greater surface area for receiving odors. They also had large spaces that could serve as resonating chambers for noisemaking. David Weishampel at Johns Hopkins University once made some pipes that you could blow into to approximate the noise the lambeosaurs might have made. He did it for a television show, and it was not meant to be highly sophisticated research, but the pipes worked and there is plenty of evidence that the lambeosaur crests would have worked this way, too. The hadrosaurs were probably also able to make noise. There is a hollow passage in the bone of the upper jaw, in a number of hadrosaurs (including *Maiasaura*), that could have worked something like a flute. The passage may have been covered or partly covered by skin.

IN THIS WORLD OF DUCKBILLS, *Maiasaura* was a terrestrial, upland dinosaur typical in some ways of the other hadrosaurs. She lived in the middle of the hadrosaur span, around 80 million years ago in the late Cretaceous period. *Maiasaura*'s most distinctive physical trait is the nature of her skull. This was, of course, what made us realize we had a new species. As I said in the last chapter, after we had finished digging out the first nest of babies in August 1978 we took the adult skull that the Brandvolds had found and brought it back to Bob's yard. We removed the cast from the skull, washed off the dirt and tried to figure out what species we had. At first I thought the skull was from a dinosaur called *Prosaurolophus*, a fairly well-known hadrosaur. *Prosaurolophus* lived in the right time and place, and its fossils show a little crest on its skull—not an elaborate one like the lambeosaurs have, just a small, kind of distinguished little fillip on the top of its skull. The skull from the Peebles ranch had just such a crest. When we were washing off the matrix (stone and dirt) from the skull with the hose, at very low pressure, the first thing we saw was that crest. I was predisposed to think that the skull was from a dinosaur that was already known, because one is more likely to find known than unknown dinosaurs. But then, as the work proceeded over the next hour or two (this gives you an idea of the rate at which we let the water dribble from the hose), I began to see from the snout that this was a different creature altogether.

The skull had a distinctive nose, different from all other hadrosaurs with the possible exception of *Telmatosaurus transylvanicus*. That dinosaur was found around 1900 by one of paleontology's most eccentric figures, Franz Nopcsa, a Transylvanian baron who was a spy, fossil hunter and itinerant European intellectual, and who ended his life by suicide. Nopcsa's *Telmatosaurus*, like the skull I was holding, had a long muzzle with a small external naris, or nose hole, and a long

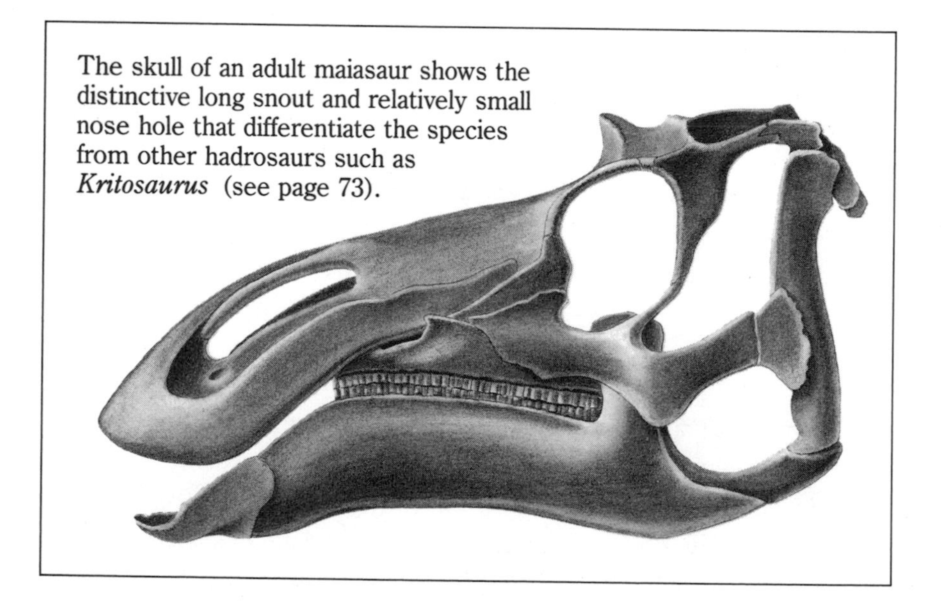
The skull of an adult maiasaur shows the distinctive long snout and relatively small nose hole that differentiate the species from other hadrosaurs such as *Kritosaurus* (see page 73).

expanse of bone between the naris and the orbit. The naris is more or less where you would expect a hole for a nose to be. The orbit is the hole for the eye. All North American hadrosaurs have, in colloquial terms, short snouts and large nose holes. This skull had a long snout and a small nose hole. That was enough to tell us that we had a new species.[8]

We finally determined that the babies were members of this same species when, during the winter of 1979, we pored through the baby bones and found a lot of fragments of baby skulls. When we compared them to the adult skull, it became clear that both sets of fossils were from the same species of hadrosaur.

The facial region of *Maiasaura*'s skull is, as one would expect, in the shape of a duck's bill. And at the end, just as at the end of a duck's bill, there were during life ramphothecae—horny growths on the upper and lower jaw. If you look carefully at a duck, you'll see that at the end of the bill is something like a pair of horny lips. These are ramphothecae, and they are usually a slightly different color from the bill. We know that *Maiasaura* had ramphothecae because the bottom and top jaws of a skull don't meet when you close them together; something else was attached to those jaws, in life. Furthermore, the ends of the upper and lower jaws are pebbly and rough, with numerous holes that would have served to allow blood flow directly to the ramphothecae.

Maiasaura's crest would have had some kind of skin or cartilage growth attached to it, so that in life this would have been some kind of display structure—*Maiasaura*'s equivalent of the colorful skin on the top of a chicken's or rooster's head. I have no idea what color *Maiasaura* was in life. Perhaps brown, or green, maybe red. The skin impressions of other duckbills show an unevenness that may have been reflected in spots, or splotches of color. The splotches could, however, have just been different shades of basic brown or green, like the variations in darkness on an alligator's back.

In terms of stature, *Maiasaura* was like the proverbial Mama

Bear. She was middle size. The average adult size of the maiasaurs we've found is close to 25 feet from nose to tail. (In describing her physical characteristics I will, in a sense, be jumping ahead of myself, making use of the many *Maiasaura* fossils we were to find on the Peebles ranch throughout our six years of digging.) I can't say that this is as big as *Maiasaura* got. Although duckbills were remarkably sophisticated reptiles, they were still reptiles. Mammals and birds have limits to their growth, which is evident in the structure of their bones. Reptiles do not stop growing. An alligator just gets bigger and bigger until it dies, although it gets bigger very slowly toward the end. The bones of dinosaurs show that they did not have any set limits to their growth, so the ultimate size of any given dinosaur was subject to diet, time and environment. We have, for instance, two very big maiasaurs that we found on the Peebles ranch, and these would have been at least 30 feet long. Probably maiasaurs didn't often get bigger than that, but I can't say for sure. That's not quite as big as it sounds. Hadrosaurs weren't built like elephants or rhinoceroses. They were more slender, or gracile. For its maximum 30 feet, a maiasaur probably weighed two or three tons, depending on how thin or full-fleshed we imagine it. A powerful draft horse, much shorter in length and height, can weigh a ton. As to how this size fits with other varieties of dinosaur, there were dinosaurs of all shapes and sizes. Some were as small as chickens. One may have been as small as a starling. And others, like the sauropods, were incomprehensibly huge. Even though two-thirds of the 75-foot-long *Apatosaurus* was neck and tail, it still weighed about 30 tons.

Maiasaura had massive hind legs and somewhat thinner forelimbs. All the duckbills had hind legs that were bigger and stronger than their forelimbs, but there was considerable variation in the size and heaviness of the bones in both sets of limbs. Some had very heavily built forelimbs, with the humerus (the upper bone) particularly massive. Most of these were lambeosaurs. *Maiasaura* was constructed in the opposite fashion. Of course, you have to remember we are talking

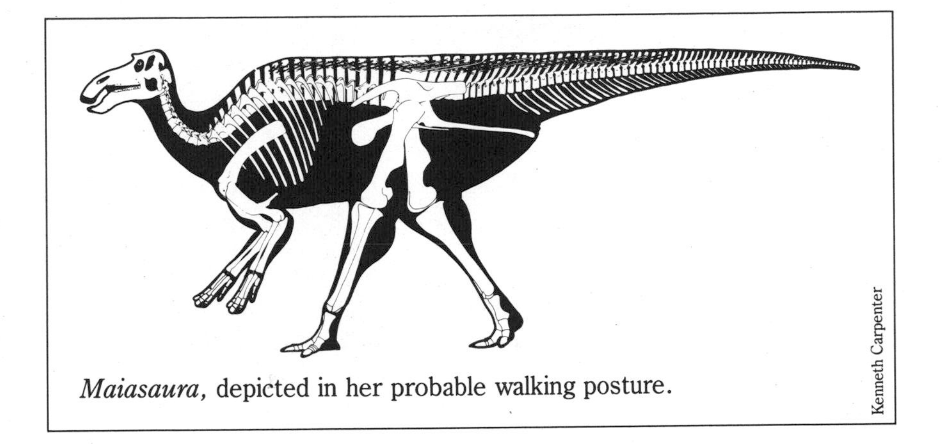

Maiasaura, depicted in her probable walking posture.

Kenneth Carpenter

about a 25- to 30-foot dinosaur. A *Maiasaura* humerus is still big—about two feet long, as thick as a two-inch pipe in spots and in other sections flat and wide. Only for a dinosaur would this be a welterweight humerus. *Maiasaura*'s hind legs were much more massive. I'm tempted to guess that this makes *Maiasaura* more bipedal than the other hadrosaurs. The ones with the longer, thicker forelimbs might more easily have gone on all fours if the occasion called for it. *Maiasaura* could not have depended too much on her forelimbs to support her weight, and the hind legs were certainly massive enough to enable her to get around quite well.

Like all other hadrosaurs, she had three toes on her hind feet and four digits on her front feet. Although we have no skin impressions preserved of *Maiasaura,* I would be willing to bet that, like some other hadrosaurs whose skin impressions have been preserved, she had a frill running down the backbone that served no purpose other than display. It may well have been more prominent on the males. I say this because I know, for reasons I'll explain later, that *Maiasaura* was a social herd animal, and modern social animals often have characteristics, like antlers, that serve only as displays to attract the attention of potential mates.

The final thing to be said about the physical *Maiasaura* has to do with her evolution. *Maiasaura* was a physically conservative, but nonetheless advanced dinosaur.[9] In her overall shape, and particularly in the contours of her face, she shows little evolutionary change from her iguanodontid ancestors. In those respects she is something like a generalist, a generalized hadrosaur. In her teeth, however, and in some other characteristics, she has changed quite a bit. Her dental battery is more elaborate than that of her ancestors, and the teeth are different in form from those of the iguanodontids and other early hadrosaurs.

Another kind of animal might show more dramatic, rapid evolutionary change in, say, the shape of its snout. A good example of this kind of animal is an as yet unnamed hadrosaur that comes from the very bottom of the Two Medicine formation. This dinosaur was found by Barnum Brown of the American Museum of Natural History in 1916 near the Two Medicine River, which gave the formation its name. He thought it was some kind of kritosaur, which is a genus of hadrosaur. I don't think he got it right, and I'm working on giving it a new name, but I haven't yet figured out what kind of dinosaur it really is. I do know that the creature is very close to its iguanodontid ancestors; it is one of the first of the hadrosaurs to appear. As might be expected, it has a lot of primitive, iguanodontid-like characteristics, such as its teeth. And yet it also has a greatly extended external naris, or nose hole, in a very noticeable arch. Perhaps for the time being we should call it hook-nose. The nose arch sounds insignificant, but it's a clue. In later deposits in the Judith River formation (which preserve the same time period as that in which *Maiasaura* lived, but in the lower rather than upper coastal plain) there is at least one whole lineage, with several genera, of hadrosaurs with those nose arches. They look very much like descendants of the original hook-nose.

These two animals, *Maiasaura* and hook-nose, seemed to me to be involved in two different evolutionary courses: one a rapid radiation of new genera and species, the other a process of slow, gradual refinement. Both courses relate to what was going on with the inland sea, which was the dominant factor in all of North American Cretaceous life. In the first chapter I described the transgressions and regressions, or expansions and contractions, of the Cretaceous sea. Each time the sea expanded, it ate up coastal plain and, in effect, destroyed habitat. Over the course of thousands, or hundreds of thousands, or millions of years, it squeezed all the animals living on those coastal plains into a smaller and smaller area until, in some of the transgressions, it reached to the mountains themselves. Just before the beginning of the Two Medicine formation, such a transgression occurred. The result was that all the species and genera that survived the loss of habitat were pushed into small pockets in the mountains.

Now, this is a recipe for rapid evolution. The work of the renowned evolutionary biologist Ernst Mayr showed that geographic isolation promotes the development of new species because a small group changes, evolutionarily, more quickly than a large population. And according to the recent punctuated equilibrium theory of how evolution occurs, proposed by the paleontologists Niles Eldredge and Stephen Jay Gould, isolation and other stresses promote periods of rapid evolution that punctuate long periods of relative stability during which there is little change. Robert Bakker has applied this theory to the movements of the Cretaceous inland sea to suggest that the transgressions caused numerous extinctions and rapid speciation.

I think *Maiasaura* and hook-nose provide evidence to support and add to these ideas. They also suggest future avenues of study. What we find in sediments that date from before the transgression of the Colorado Sea are iguanodontids. As the transgression occurred, the growing sea wiped out habitat and squeezed everybody into the mountains. We don't have a good record of what went on during this time, but we do have the Two Medicine formation from the period when the sea began to recede—and we find in it not iguanodontids but hadrosaurs.

Another kind of evolutionary pressure had been introduced when the sea receded. Suddenly vast new territories, new ecological niches, were opened up to be colonized by opportunistic species. With all this open space, another period of rapid speciation may have occurred. Hook-nose and *Maiasaura* seem to fit into this scheme differently. With hook-nose we can see good evidence of rapid radiation of descendant species as the coastal plain opened up. All hook-nose's evolutionary grandchildren are out there with a whole spectrum of elaborated nasal arches. They'd been evolving up a storm. But *Maiasaura* is from the same generation as these grandchildren, and she is very little different from her ancestors. It seems that she proceeded quietly, evolving gradually, refining certain important characteristics such as teeth.

The difference, I think, lay in geography. I suspect that *Maiasaura*'s ancestor appeared at the same time that hook-nose emerged, both in the mountains. When the new coastal plain opened up, however, the two ancestors took different directions. While hook-nose colonized the new territory and spawned opportunistic descendants, the maiasaur ancestor stayed near the mountains on the upper part of the coastal plain. This was not as lush a territory. It may have had fewer niches, with fewer opportunities for radiation, and instead of providing the staging ground for numerous new species to emerge, it provided a testing ground to improve the ones that stayed there.

This is the kind of idea that one can test in paleontology, but not on the Peebles ranch. This land contains sediments from only the upper middle part of the Two Medicine formation. I needed to go to the bottom and look for a contemporary of hook-nose that could have been the ancestor of *Maiasaura*. And I needed to look a little bit higher in the formation for creatures intermediate between hook-nose and his presumed descendants. Such finds would provide examples of a transitional species, showing the process of evolution, something that had not been done, in this kind of detail, with dinosaurs.

This was one of the ideas generated by the finds on the Peebles ranch, and eventually I followed it up. But that was years in the future. My ideas about *Maiasaura*'s evolution did not even begin to take shape until we had been digging up fossils from the Peebles ranch for several years. Where I left off in the story of the dig itself was the summer of 1978. Bob and I had found one nest and a skull. We had another six years of discoveries to go.

CHAPTER 4

NESTING IN COLONIES

I've delayed long enough in introducing the Willow Creek anticline. So far, I've said only that we found the babies on the Peebles ranch in the Two Medicine formation. Well, within a geological formation the rock beds are often bent into twists, hills, valleys and folds, and the vocabulary of geology has numerous terms to categorize these structures or features. One of these terms is "anticline." The geological feature in which we found the first nest is an anticline named in traditional fashion after a local watercourse, Willow Creek. This is a good-size stream when it's not dry, with some trout in it and banks covered with just the sort of tangled willow thickets that grizzly bears love. That first nest and all our later discoveries came from the anticline itself, or just off the edge of the anticline.

An anticline is a wrinkle, or fold, in the earth—a little hill. In talking about the growth of the Rocky Mountains, I described how the thrusting of the mountains created a foredeep, or dip, in front of them. Similar processes account for the anticline, but on a smaller scale. The

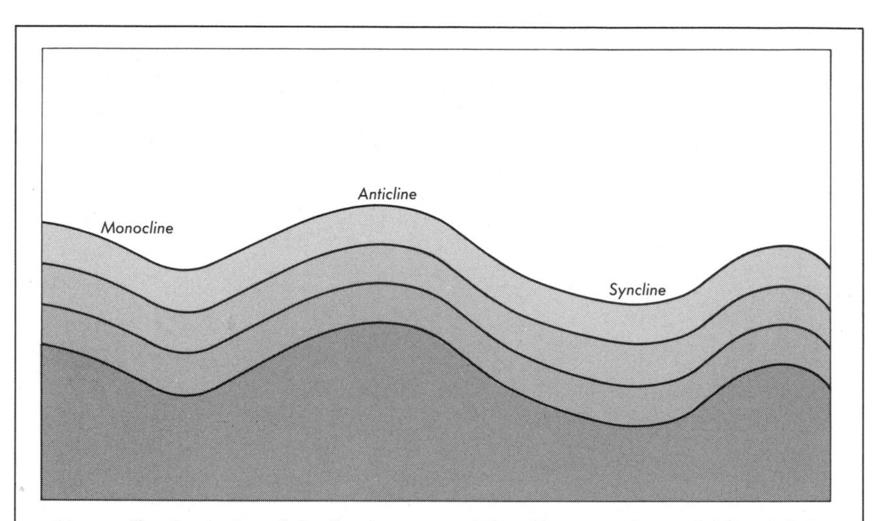

Above: Geological activity in the area of the dig caused a wrinkling of the rock beds laid down earlier by sedimentation. An anticline is a kind of hill; a syncline, a dip or valley; a monocline, a single slope.

Below: A cross section of the anticline outlines its shape before erosion. Various fossil sites (some of them mentioned in later chapters) are indicated on their different levels, or horizons.

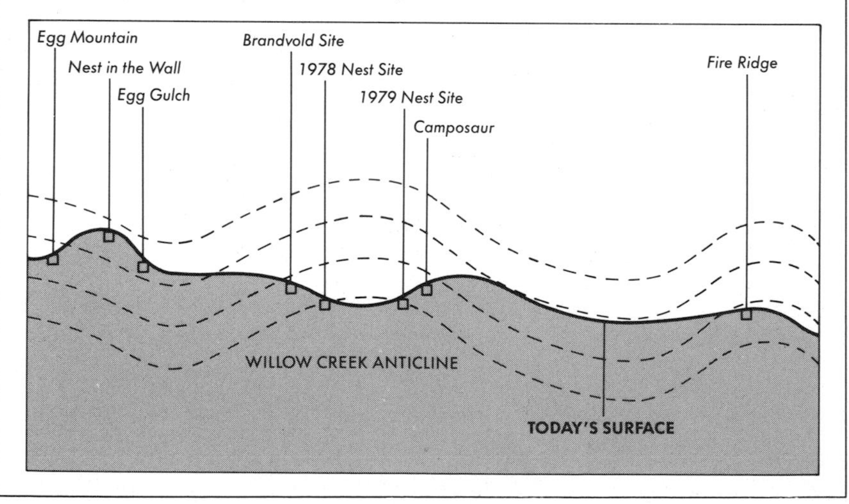

foredeep might have extended 100 or 200 miles in front of the mountains. The anticline proper is only about one square mile. The entire area that we covered in the dig as time went on was only two square miles. Still, the principle is the same. When geological forces such as the movement of mountains or earthquakes or volcanic activity cause the skin of the earth to bend and break, many small dips and hills are formed. A dip, or concavity, is called a syncline. A hill is an anticline. One slope of a hill (when the rest is missing) is called a monocline.

The hard thing for someone not versed in geology to grasp is that you can't see these hills on the surface. Aboveground, the Willow Creek anticline is just a piece of sorry pasture land with a lot of scarring and erosion, hills and gullies. There are areas of barren green mudstone and reddish sandstone. Where there is enough dirt to support plants, you find short grass, prickly pear and wild honeysuckle. There are black widow spiders, a few scorpions, Richardson's ground squirrels, and badgers, which eat the ground squirrels. None of this, and particularly not the hills and rises that you have to walk over, bears any necessary relation to the shape of the actual anticline. The surface is made up of temporary accumulations of dirt, momentary features of the landscape. They may be gone in a hundred or a thousand years. Geologically, they are no more significant than the cracks and flakes of a bad paint job on a clapboard house. The structure of the house is hidden. And that's what the anticline is, part of the underlying structure of the land.

To find the anticline you must locate the layers of rock underneath the surface, which are exposed in some places through erosion, and follow these layers to see what transformations they have undergone. You must determine how they are shaped and how they are tilted. The Willow Creek anticline is not easy to follow. I sought out two graduate students in geology, Will Gavin and John Lorenz, to study the geology of the Two Medicine formation and the Willow Creek anticline to give us a kind of geological map of the territory.[1] They found that after the rock beds had buckled to form an anticline, they had also been bent and twisted so that the hill forming the anticline now curves around in a crescent shape and is in places so jumbled that it's all but impossible to figure out which rocks were originally on the bottom and which on top. Furthermore, the whole top of the anticline has been scooped out by erosion.

The anticline is about 160 feet thick, and it has seven distinct, fossil-rich rock layers. Just as the Two Medicine formation is broken up into large layers of shale and mudstone, the Willow Creek anticline, which is just a thin slice at the top of the Two Medicine formation, dating from about 80 million years ago, is itself sliced into these much thinner layers, each of which has characteristic rock and fossils. The layers are called fossil horizons. One horizon, for example, shows a lot of small streams, and another a large lake, each with a different kind of dinosaur represented in its fossils. Like the larger layers of a geological formation, the horizons of the anticline show a progression in time, with the bottom horizon being the oldest and the top horizon the most recent. It was from a spot on the bottom horizon, in the scooped-out center of the anticline, on a pebble-strewn nub of dirt and rock, that Bob and I dug up the first nest of baby maiasaurs. And it was on this same horizon that, over the next few years, we were able to find eight more maiasaur nests.

THE SUMMER OF 1979 was our first full field season. We camped that year not on the Peebles ranch, but on the banks of the Teton River on land owned by A. B. Guthrie, the Pulitzer Prize-winning novelist who wrote *The Big Sky.* We got to know Guthrie, who's usually called Bud, in a roundabout way.

The town nearest the Peebles ranch is Choteau, which has a campground, restaurants, clothing stores, and gift shops for tourists.

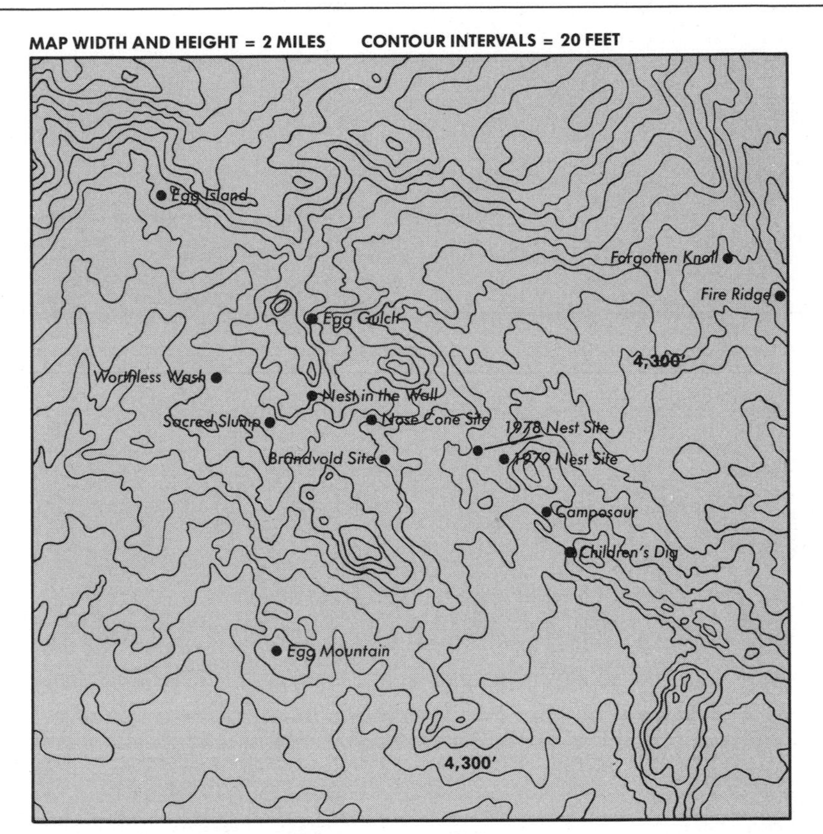

A topographical map of the Willow Creek anticline and nearby area shows the surface distribution of the dig's various fossil sites. The 1978 and 1979 nest sites mark the first maiasaur nesting ground.

It has a population of only about 2,000 people, but it looks bigger because it serves a large surrounding area of ranches and farms. During the time Bob and I were waiting for the Brandvolds to get us permission to dig, we explored all that Choteau had to offer, pretty quickly. One of the spots that was of particular interest to us was a rock shop and tourist museum in town (it's now owned by the Brandvolds) that had gems, Indian artifacts, and dinosaur fossils. There was a young woman working at the store who was pretty, shapely, and open and friendly in the way the rural Westerners tend to be. It was almost impossible not to strike up a conversation, and naturally what we started talking about was why we were in town. Her name was Amy Luthin and she was Bud Guthrie's stepdaughter. It was through Amy that we met Guthrie, and then the Peebles. And because of the meeting we ended up getting direct permission from the Peebles to dig on their land. Bud Guthrie let us set up camp on his land about a 10-minute drive along a good gravel road from the site where we were going to dig, the spot where we had dug up the first nest.

In planning for that 1979 season, we needed to make a big jump in size and organization from the previous summer. We were going from two guys in a van to a season-long, full-fledged paleontological dig. The key ingredient for such a change was money. We had to have tools and materials to prepare the fossils. We had to pay for transportation, for food for ourselves and the volunteers, and for at least a nominal salary for Bob. (I was being paid by Princeton as a preparator, so I didn't need extra money for myself.) I figured that we needed $10,000, and I set out that winter to get it. Since Princeton was not immediately forthcoming with the money, I tried a more unconventional source. I wrote to the Rainier Beer company in Seattle, Washington, explaining that Bob and I drank a lot of Rainier whenever we were in the field. (This was absolutely true then and continued to be true as the dig grew. Depending on the size of the crew at the anticline in later years, our summer supply would range from 50 to 100 cases. Bob and I made a point of putting together a beer kitty, which we kept separate from the other accounts; we did not think most of the funding agencies would consider beer an appropriate expense.) We told Rainier that if they would support the dig, we'd be happy to acknowledge them at the end of our papers, the way scientists always acknowledge the organization that provides the grant. I think Rainier must have contacted Princeton, because the chairman of the geology department (which vertebrate paleontology was a part of) informed me that the process of soliciting grant money from corporations was conducted on a university-wide basis. Free-lancing was not allowed. Shortly after this conversation, Princeton found the $10,000.

That first year at the riverbank on Bud Guthrie's land, we had 10 tents and Bob Makela's teepee. The teepee was a large affair, canvas stretched over 20-foot pine poles and the interior strewn with bits of carpet and pillows. Later I came to use one myself, and I got a small one for my son Jason, who came out to the dig each summer. We set up another one for visitors, and some volunteers brought their own. Teepees became the official domicile of the Willow Creek anticline dinosaur dig, with good reason. They're the most comfortable way to live on this land, perfected over centuries by the plains Indians. They're airy, full of light, and sturdy. You can build a fire in them, and they'll stand up to the 80-mile-an-hour winds that sometimes roar down from the Rockies and tear up tents. The teepees we used were all modeled after the original Blackfoot teepees, since the dig was in Blackfoot country. These teepees have a basic structure made of four poles leaned against each other with their tops interlocking. The four poles give the teepee its strength. On them other poles are laid on which the canvas, or in the old days buffalo hide, is stretched. The Sioux used a tripod as their basic structure, and their teepees also have a slightly different shape, not quite as tall, and with more of a slope.

The crew numbered 13 that year, including a full-time cook and Bob, both of whom had paid, official positions. Everybody else was a volunteer, which was the way we've run all the digs since, with Bob and one or perhaps two other people drawing a small salary from the grant money. (In fact, this is the only possible way to run paleontological digs, because there just isn't enough money around to pay a full crew.) Our first task that summer was to look for nests. We were fairly certain we would find at least one more. Shortly after I returned to Princeton the previous winter, I'd received a package from Amy Luthin. She'd been out to the site of the first nest during the time Bob and I were digging it up, and later in the summer she went out herself to poke around. She found something on the surface that looked interesting and mailed the pieces to me. Apart from this surface collecting, she left the site undisturbed.

In the package from Amy were fragments of eggshell and bone that looked like they might have come from hadrosaurs even smaller than the ones we had already found. This material was not from the same spot as the first nest, but the location was on the same level of the anticline as the first nest. A concentration of bone and eggshell like that suggested to me a second nest. But when we set out in July 1979 to uncover what Amy had found, we couldn't discern the outline of a nest; we couldn't tell where to dig and where not to dig. So we didn't dig at all. We collected from the surface and let the site weather for a year. This is a fairly common practice. Wind and rain are not only inexpensive; in the short term, they have a more delicate touch than a paleontologist's ice pick. (And when I say we let the site weather, that doesn't mean we sat around and waited. We went on to prospect for other nests, to find them, to dig them up, and to dig up a variety of other fossils. We always had several bone deposits that we were excavating at any one time at the Willow Creek anticline.) The next year, we again collected from the surface and again we swept the loose dirt and screened it for bones. Again we decided to let it weather. Not until 1981 did we actually start to dig into the mudstone. Then we found that the reason it had been hard to see the extent of the nest was that the nest was not complete. What remained was only a portion of the bowl-like shape we had seen in the first nest. But that portion did contain the badly weathered bones of seven extremely young dinosaurs and one embryo.

Amy Luthin's nest, and even in 1979 we thought it would turn out to be a nest, was significant because it was the first hint that

The shortgrass prairie and eroded badlands of the Willow Creek anticline; to the west, the Rocky Mountains rise in the background. Camp (*right*) was a collection of teepees, tents and trailers.

Museum of the Rockies Photo

Workers. *Clockwise, from left:* Crew members sift for eggshell and bone; the 1982 crew poses under the kitchen tarpaulin; bones are uncovered at the Brandvold site.

In the characteristic paleontological stoop, crew members search the chipped rock of Egg Mountain for fragments of eggshell and bone.

Joe McNally Photo

Regulars. *Counterclockwise, from top:* Jason Horner, inspecting the top of Egg Mountain; Bob Makela and Jill Peterson, putting a plaster jacket on two legbones at the Brandvold site; Bob Makela.

Museum of the Rockies Photo

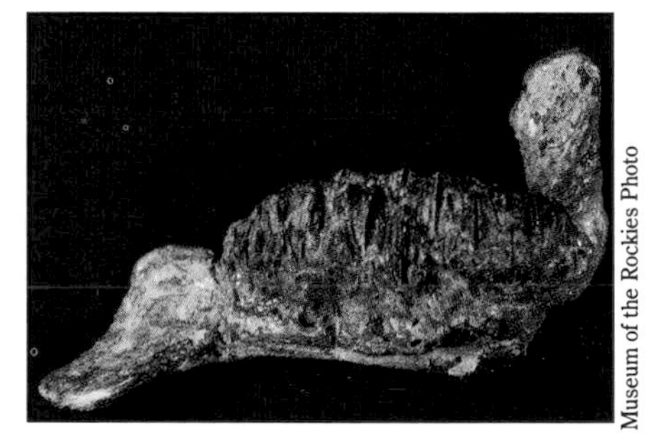

Museum of the Rockies Photo

Joe McNally Photo

The Good Mother Lizard. *Top:* The reconstructed skull of the original specimen of *Maiasaura peeblesorum,* found by Laurie Trexler and prepared by Pat Leiggi. *Above:* The jaw of a nestling maiasaur, two and a half inches in length. *Right:* An ice pick is one of the standard tools, used to chip away dirt and fossil bone; here, work is done on a piece of legbone still embedded in the ground.

The paleontologist (*right*), contemplating his fossils; in front of him lies a thighbone from the Brandvold site that has been cast in a plaster jacket and pulled from the ground. *Below:* Barbara Haulenbeek packs another jacketed bone from the same site back to camp. In the winter, the protective jackets will be removed so that the bones can be carefully cleaned, studied and, if necessary, reconstructed.

Joe McNally Photo

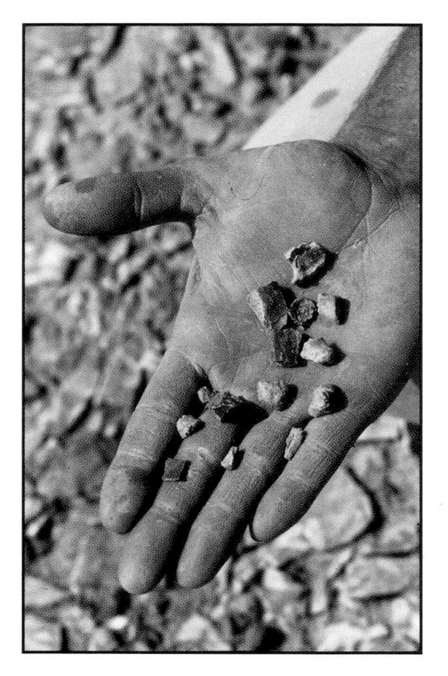

Small treasures. Fragments of bone (*left*) from a nest of hatchling maiasaurs look like so many dull bits of rock. The cross section of a hypsilophodontid egg (*below*) shows the hints of bone that signal the presence of a fossilized embryo. Egg Island yielded 19 eggs containing such fossils, the first preserved embryonic dinosaurs ever found.

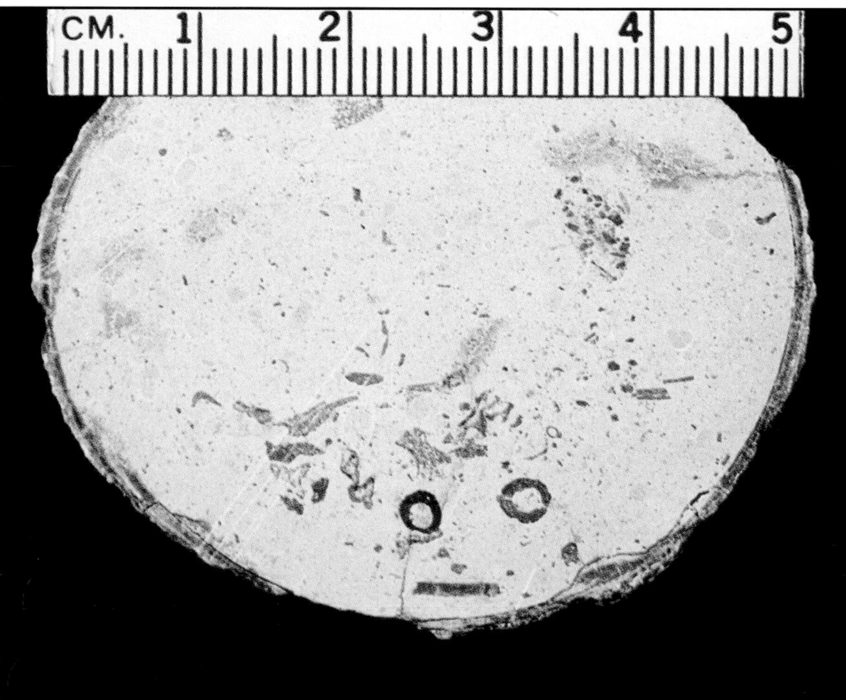

Museum of the Rockies Photo

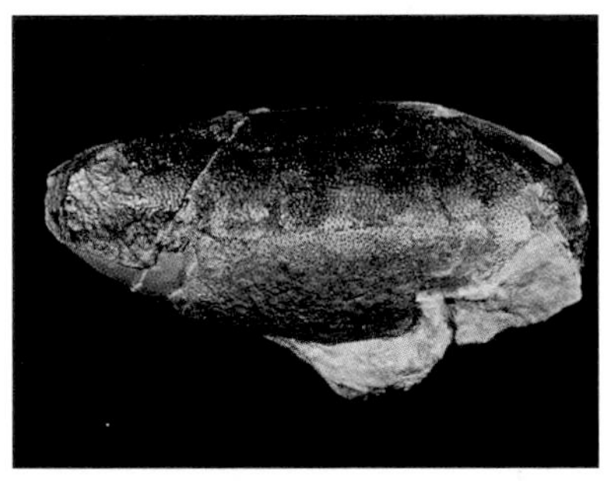

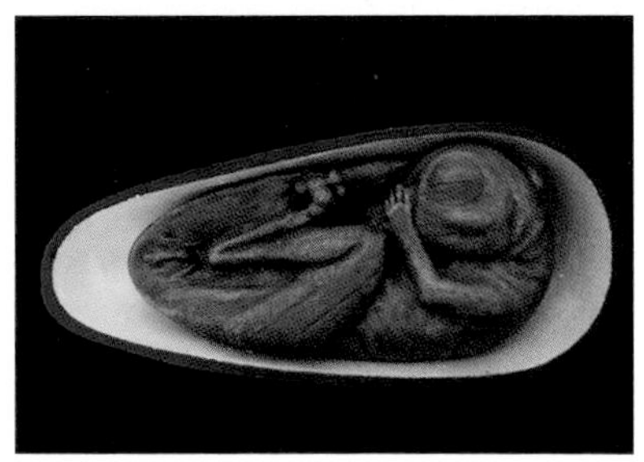

A reconstructed clutch of hypsilophodontid eggs from Egg Mountain (*above*), laid by the mother dinosaur with their ends embedded in soil; these eggs, preserved whole, never hatched. *Right:* An egg of an unidentified variety of dinosaur, also from Egg Mountain, and Matt Smith's model (based on a CAT scan) of an embryonic hypsilophodontid reaching hatching age.

Museum of the Rockies Photos

After a failed attempt to lift a jacketed maiasaur nest out of Egg Gulch by helicopter, Bob Makela, Lisa Ulberg and Robin Voges cut the nest in two. The second try was successful.

Jack Horner walks over the Camposaur pit, one part of the big bone bed that contained the fossil remnants of a herd of 10,000 maiasaurs.

Marion Brandvold's find was not an isolated one—that more than one dinosaur might have come to this upland plain to lay her eggs. Amy's find was also intriguing because it was on the other side of the scooped-out center of the anticline from the first nest. And even though the middle of the anticline was missing, we could tell that the rock in which both these nests were located was the same rock. They were both on the same bottom layer of the anticline, on what is called a single fossil horizon. The layer was of mudstone laid down by stream flooding. And, having the signs of two nests in that layer, it behooved us to look for more. We did that by following the mudstone layer all over the anticline, on our hands and knees.

Up and down the anticline we went, day after day, seven or eight people, some of us wearing rug-layer's knee pads, crawling on sharp rocks and pebbles in the sun, grousing and moaning about either the rocks or the knee pads themselves, squinting at the ground two or three feet away to try to distinguish in the glare of the sun a half-inch bit of dusty black rock that might be a fossil bone from the other half-inch bits of dusty gray, brown, red, green and white rock.

The normal way to look for dinosaur fossils is just to walk along and look at the ground. In the Judith River formation, which preserves swampy, lowland terrain, you step on fossils all the time, and you can see them when you're standing up. In the preserved dry uplands (dry then and dry now) of the Two Medicine formation, on the scruffy slopes of the Willow Creek anticline, there is no such luck. You could walk the terrain for a year and not see anything. You have to keep your nose somewhere between a foot and 18 inches away from the ground in order to have a chance of finding anything. In fact, because of the difficulty of work at the Willow Creek anticline, I tried to get volunteers who had not been spoiled by other, easier dinosaur digs. After two or three days or a week of crawling and not finding anything good, experienced paleontological volunteers might well have packed their bags and left, discouraged and dispirited. Newcomers didn't know any better. Once the newspaper and magazine articles and the television reports about our finds started to appear, right after we found the first nest in 1978, I began to get letters from people who wanted to volunteer. There were also people at Princeton who wanted to help, as well as people whom Bob and I knew in Montana. For unknown volunteers, I usually waited until they had not only written two or three times, but had also called me on the telephone. At that point I would explain how difficult it was, that there wasn't even a latrine (we just carried a shovel over a hill), and that the routine was six days of work and one day in town to buy food and wash up. After all this, if the people I was doing my best to discourage were still interested, I gave them directions and told them to bring a tent or a teepee.

My guess, and I was usually right, was that the ones who were persistent enough to get me to accept them would be persistent enough to get through the hard work and the discomfort to experience the rewards. The heat and the bruised knees faded in importance as soon as we found something. Some days, of course, were better than others. The entries in my field journal for July 1, 1979, show that we found the remains of three nests in that one day. On July 1, our knees didn't hurt at all.

For each of the nests we found that day, the signal that we had something was a concentration of crushed eggshell on the surface. Once we had found an eggshell concentration, we got down on our hands and knees with ice picks, whisk brooms and dustpans. We carefully removed fossils and dirt, digging down to look for the telltale bowl-shaped boundary between green and red mudstone that marked a nest. We screened the dirt we took from each site for fossils, mapped each nest on a grid and made a map of all the nest sites. The concentrations of eggshell did not always turn out to mark a nest; sometimes we would dig down and find no bowl-shaped boundary, no outline of a nest. Eggshell on the surface was not always a reliable indicator of a nest because the Willow Creek anticline was unlike most

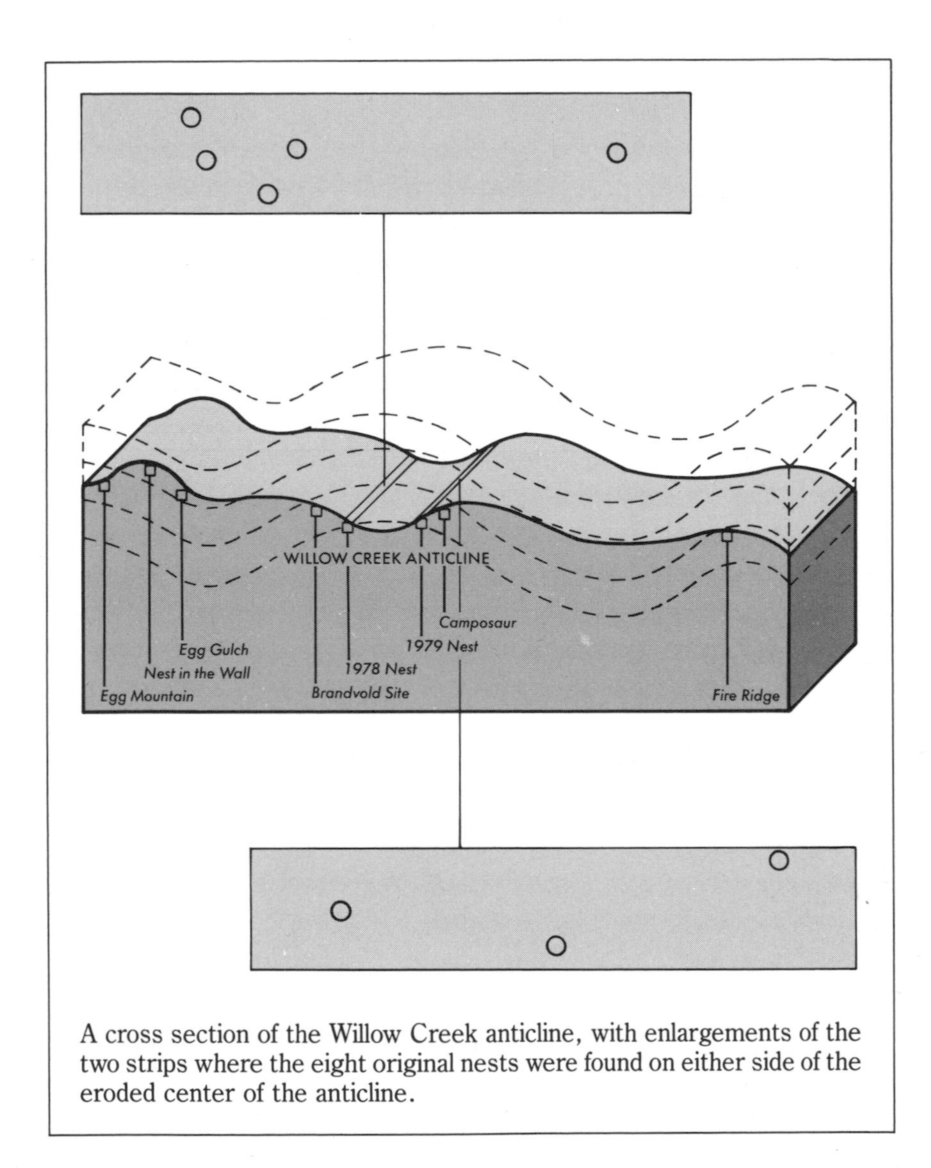

A cross section of the Willow Creek anticline, with enlargements of the two strips where the eight original nests were found on either side of the eroded center of the anticline.

of the paleontological digs in the world, where dinosaur eggshell would be a wonderfully rare find. At the Willow Creek anticline, eggshell crumbs were the most common variety of fossil. After the dig got going, we would often give dinosaur eggshell away to visitors as fossil souvenirs.

What distinguished a nest from a simple concentration of eggshell was first of all that demarcation between the green and red mudstone showing that a hollow bowl had been dug in which to deposit eggs, and second a very heavy concentration of eggshell *under* the surface—in the bowl. We didn't find intact eggs at all, just the broken fragments. That made sense, given our notion that these creatures stayed in the nest after they hatched and therefore trampled underfoot the eggs they had come from.

We kept searching that one fossil horizon through the summer of 1980 and then through the summer of 1981, alternating the work there with other sites where we were finding other fossils. Eight times the eggshell led us to uncover a nest—eight times on the one horizon. In two of those nests there were bones of babies as well as eggshell. But we never found intact or even partially intact eggs there. By the end of 1981 we had found, dug up and marked the location of all eight nests, all of them in this little dip in the anticline, a miniature badland of dull, brownish reds and gray-greens, of mudstone flats and blocks of sandstone that stood as hills or small-scale bluffs. There was next to no vegetation, and the signs of erosion—water-cut channels and crumbling rock—were everywhere. On this harsh and unwelcoming terrain, we spray-painted eight blaze-orange circles, each of which marked a nest. To me those eight circles made an extraordinarily powerful image because they did not represent just eight nests made at random over the course of decades or hundreds or thousands of years. Instead the peculiar geology of the anticline allowed us to pinpoint the nests in time. We believed that they had all been laid in the same year, the same nesting season. They were no random collection. Our conclusion was that at least eight *Maiasaura peeblesorum* females had gathered together, spacing out their nests, and laid their eggs and raised their young in a colony reminiscent of penguins nesting on the coast of

Antarctica. Nothing like this had ever been uncovered before. Here were dinosaurs—the beasts that traditionally had been depicted as nothing more than big lizards, slow, stupid and thoroughly reptilian—acting like the most social of birds. And here was a snapshot of a fleeting moment—one year—that occurred 80 million years before we had come to sweep, dig, and brush away the dirt and rocks so that we could spy on the lives of dinosaurs.

IT'S A BIG LEAP from eight spray-painted circles to the scene I want to depict: the nesting ground I believe was there. In order for this scene to be accurate, for my conclusions to be valid, the nests had to have been made in the same geological instant—a single year. If they had been made at different times, perhaps by the same dinosaur returning to the same general area, there was no nesting ground. In most circumstances, dating 80-million-year-old sediments with this kind of precision is impossible. But at the Willow Creek anticline we had a bonus in the form of marble-size nodules of calcium carbonate called caliche.

These nodules made crawling around in the Willow Creek anticline excruciatingly painful. Even with knee pads, the caliche pebbles dug into our kneecaps. They also made walking dangerous. Although the nodules were sharp-edged, they were round enough to roll. It doesn't matter what kind of sole you have on your shoes or boots—if you don't take extreme care walking down a caliche-scattered slope, you fall. Nobody manages to take extreme care at every moment, so everybody falls. And it hurts. Your feet fly out from under you, and you land seat-first on a hillside of sharp marbles.

Caliche does not form on the surface, although it may end up there. It forms under the surface. How far under depends on how wet the climate is: the more rainfall, the closer to the surface it lies. But any given layer of caliche always forms at a specific distance from the soil surface. The way it forms is that mineral-rich groundwater fluctuates in level with the wet and dry seasons. Each time it rises, the water deposits calcium carbonate in the soil. The calcium carbonate accretes in nodules, and these nodules are what we call caliche. Caliche has been studied so much that it is known how many wet and dry cycles (or years) it takes to make a certain-size nodule. When you find a layer of caliche, you can refer to well-established tables that tell you how long the nodules took to form.[2] That, in turn, tells you how long the soil surface above it existed, without eroding or being buried by silt. Were the level of the surface to change, a new layer of caliche would form.

All the hadrosaur nests I have mentioned are about eight inches above the same caliche layer, all on the same preserved soil surface, or horizon. And the caliche nodules show that they took 5 to 10 years to form. This means that the soil horizon we found the nests on existed for at most 10 years. When you're dealing with 80-million-year-old sediments, this is really pinpointing fossil deposits in a geological instant. But I think we can go even further. I think we can argue, strongly, that those nests were all made, and the babies hatched, in one year, one nesting season.

My reasoning is that, first, we found no overlapping of nests. That's negative evidence. But if those nests came from several different years, and not from the same single year, we might have found some of them overlapping others. Certainly, I think we would have found less uniform spacing. The sites of those 6-foot-wide nests were equally spaced, 23 feet apart. Twenty-three feet is the average length of adult maiasaurs at the Willow Creek anticline. Among ground-nesting birds now, it's common for nests to be separated by the length of an adult bird's outstretched wings. This seems to allow for the maximum of togetherness and the minimum of interference. If the maiasaurs were behaving like birds, as it seems they were, this means that they not only picked the same place to lay eggs but also had a social system of sorts, with what you might call a prescribed "personal space." No doubt it was based on instinct, and not learned manners, but still it's extraordinary.

My conclusion was that those nests were definitely in their current arrangement when they had babies in them, 80 million years ago. When we stand on a slope of the Willow Creek anticline, we're not just looking at spray paint. We're looking at a dinosaur's colonial nesting ground.

This was the first site of its kind ever documented. The Mongolian find, which included numbers of nests in the same surface area, may also have been a colonial nesting ground. In fact, I'll bet it was, because after working on the Willow Creek anticline I suspect that a lot of dinosaurs nested colonially. However, the claim was never made for those nests, and the site study necessary to determine if those nests were on one time horizon was just not done.

The maiasaur nesting ground probably had more than eight nests. Remember, the center of the anticline, between Marion Brandvold's nest and Amy Luthin's nest, was scooped out. We found nests on both sides of this scoop, but the middle was missing. If there were nests in this missing section, spaced the same way, then the nesting ground would have covered at least two and a half acres and would have included 40 nests.

Later we uncovered at least two more maiasaur nesting grounds. The first one was located on the lowest fossil horizon in the anticline. At higher, or later, levels we found more eggshell, nests and babies. In each case there were several nests on what we judged (again because of caliche) to be a pinpointed time horizon of at most a few years, always spaced the same way. I don't know how much later these other nests were made, because it's impossible to know exactly what the rates of deposition of sediment were. Furthermore, there is always the possibility that erosion took place between the episodes of deposition. The different nesting grounds are separated by vertical gaps of about 2 feet to 10 feet, and if I had to guess I would say this means they were inhabited during periods separated by anywhere from tens of years to thousands of years. These are very short time periods for geology, but long enough for dinosaurs. Maiasaurs probably nested in this area for generations.

In 1982, the year after we had finished the work on the first nesting ground, we found our second nesting horizon, about 50 feet above the first one and about a half-mile away. Phil Currie from the Tyrrell Museum in Drumheller, Alberta, had come for a visit. Currie was interested in searching for nests in similar formations in Alberta, so he'd come to inspect the terrain and see what eggshell concentrations looked like. I took him to a spot where I thought we might find the kinds of eggshell concentrations that often signal a nest. We had picked it out as a good location for further exploration, but as yet we hadn't gone over it carefully. Phil got lucky. He found the heaviest concentration of eggshell I had yet seen. When we swept, screened, mapped and excavated the site, we dug out six intact lower halves of eggs, with a lot of associated eggshell. This was our first maiasaur nest that included even partial eggs. I think these eggs had never hatched, that they had been fossilized whole. The reason we found only the bottom halves was that the fossilized tops had been exposed to weather over the seasons before we had begun to work the anticline and they had eroded. Thirty feet away, also badly weathered, we found four more eggs. And 30 feet from that point we found the remains of three baby maiasaurs, each about 14 inches long. When we excavated these sites, we found that both the egg sites had the by now characteristic signs of a nest: the bowl shape and the two shades of mudstone. (This was not the case for the three 14-inch maiasaurs; there was no clear nest associated with them.)

It was this horizon that gave us our first good picture of a maiasaur egg. Here we could see at least half the egg still intact, and there were enough large pieces to reconstruct whole eggs. They were about eight inches long, oval, with a rough, ridged surface.

We kept prospecting throughout the area of the Willow Creek anticline, always looking, in one rock layer or another, for signs of nests, and in 1983, the year after Phil Currie's discovery, we found two more nesting sites. These sites gave us our first picture of how the maiasaur eggs were arranged in nests. One of them is called Egg Gulch or Egg Dance Coulee. (The latter name gives some idea of how we felt when we found it, although for some reason the more mundane name, Egg Gulch, became the one we used—probably because it was shorter.) We found two maiasaur nests there and a whole lot of other eggs and bones. In fact, every time we tried to get one fossil out, or a bunch of fossils, we ran into others. I dug through one of the nests inch by inch to see its arrangement. There were 14 whole eggs, crushed but in place. (Geological processes, not hatched babies, had done the crushing.) They were in circles, and in two layers, with eggs on top of other eggs.

The other nest at this site had 18 eggs in it, also neatly arranged in circles, also relatively intact. What we did with that nest was to dig a trench around it, in which we found the jaw of a carnosaur so small it might have been an embryo, some smooth eggs of unknown origin, and a tiny little egg about two inches long that may be a fossil bird egg. When we had the trench dug we made a plaster jacket around the nest, creating a cast five feet long, four feet wide, and three or four feet deep. It took us two years to make this trench. Along the way, we kept having to stop because we were finding fossils around the nest. We also had to build a wood platform over the nest itself so that we could dig without stepping all over it. Once the plaster jacket was done, we chipped away beneath the nest, creating a pedestal, so that what we had looked like a giant plaster mushroom. Then we broke the stem, rolled the top over into a net and called in the helicopter.

The helicopter belonged to a seismic crew that was working for Amoco, looking for oil nearby. The pilots would fly out to visit us occasionally, and they had agreed to take the nest out for us. The only problem was, it turned out that the helicopter couldn't lift the nest, which weighed about two tons. The way Egg Gulch is shaped, the wind from the propellers just shot down the gulch and blew away. There was no flat surface that would stop the air and give the helicopter enough lift to get the nest up. So we cut the plaster-jacketed nest in two with chisels, and the helicopter was able to lift off the halves one by one.

That same season, on the same horizon, only about a quarter-mile away, we found the last of the maiasaur nest sites. I don't know if it was part of the same nesting ground as Egg Gulch or not, because it's so far away, but both nest sites are from the same nesting season. That last site was found by Jill Peterson, and we called it the Nest in the Wall. It had two nests, some embryonic bones, and beautifully preserved parts of five little maiasaurs, all about three feet long.

This was quite a tally of *Maiasaura* nests, eggs and young. From 1978 through 1983 we found 14 nests, 42 eggs, at least 3 nesting grounds and 31 babies. It was enough to give us a picture of a kind that had never really existed before, of dinosaurs nesting in colonies and caring for their young. And we had so many of these fossils that we could re-create the day-to-day life of these dinosaurs. We had the material to say what the nesting grounds were like, what the environment was, to talk in some detail about the behavior of the young and adult dinosaurs. I don't think any other paleontological site up to that time had yielded as much detail about one species of dinosaur, and it was that detail that enabled us to take the unprecedented step of claiming to have evidence of how these dinosaurs behaved.

TO EXPLAIN THIS BEHAVIOR, and the evidence for our description of it, I want to concentrate on the first nesting ground: the one with the eight nests, the painted orange circles that I described earlier. This is the nesting ground we have the most information for.[3] In the area where we found these first eight nests, there are three kinds of stone:

mudstone, left by overflowing streams; caliche, which I described earlier; and sandstone. Some of the sandstone deposits are greenish and thick, and now partly eroded. They form big chunks that you can walk up and down—little hills and tufts. Other deposits of brownish sandstone are much thinner but spread over a wider area. These two different sandstone deposits represent two different kinds of streams. And they are found on different horizons (different times) in the anticline. At the level of the nesting ground, we find the thick deposits. They were left by anastomosing streams, ones that stayed within set banks, even though within those banks the channel and course of the stream may have changed often. Between those banks the sand in the stream bed piled up and up over the years. The other, thinner deposits, which we don't find at the level of the nests, were left by what are called braided streams, which wandered about in no set banks.

An anastomosing stream flowed very near the nesting ground. It was probably the stream that deposited the mud that buried and preserved the nests. Its banks were probably covered with heavy vegetation, mostly trees. We know that the stream did not support a lot of life. We have found some turtle fossils, and teeth of crocodilians (probably washed in from somewhere else), but that's about it. The stream was probably too choked with silt to have fish in it.

The nesting ground itself probably was trampled and dusty, or muddy when it rained. From the shape of the nests, we know that the maiasaurs made mounds and then hollowed them out to make a place for the eggs. I suspect that the mothers may have dug with their powerful hind legs and shaped the mounds with their forelegs. I say the mother simply because it's more common among reptiles and other animals for the female to be the one who makes the nest and cares for the young. It's not, however, universal. The fathers do the child care, if you want to call it that, in some species of fish and birds (to take two examples), and it's possible that it was the maiasaur males who made the nests. Still, the odds are against it. In any case we can't tell the

A colony of *Maiasaura peeblesorum* gathers, perhaps in the spring of the year,

to build nests, lay eggs and tend the young.

difference, on the basis of the skeletons we have, between male and female maiasaurs, so it would be impossible for us to make a firm conclusion on who dug the nests.

Assuming it was the mother, her next step would have been to squat over the mound, using her forelimbs to hold herself steady, and deposit her eggs. The eggs themselves were hard like birds' eggs, not leathery like snakes' eggs. They were narrow ovals and were arranged in the nest vertically, in circles. And, as I mentioned, for some reason that is beyond me the dinosaurs seem to have laid the eggs in two layers, one on top of the other. I imagine the mother dinosaur brought vegetation to cover her nest and keep the eggs warm until they hatched. I'm sure the mother did not sit on the eggs. She was just too big.

What happened after hatching is the most interesting part. To re-create this stage of the nesting drama, we look primarily to the fossils themselves. All the young in that first nesting ground, except one isolated skeleton of a juvenile, were taken from two nests: the one Marion Brandvold found and the one Amy Luthin found. I described earlier the babies that we found in Marion Brandvold's nest, which were about three feet long and had well-worn teeth. Amy Luthin's nest, the one that she found at the end of 1978 and that we finally dug out in 1981, had babies of a different age and size. Out of Amy Luthin's nest came a jumble of legbones, the odd piece of skull and jaw, and a few other bones. These bones showed us that the babies in this nest were only about 14 inches long, roughly half the size of the babies in Marion Brandvold's nest. They could not have been out of the egg very long, because in one jaw you could see that some teeth were worn and some were not. Having some teeth that showed no wear at all had to mean that these dinosaurs were extremely young. They might have lived a week out of the egg, but not much more. So this was the size of the maiasaurs when they hatched. And we knew that they grew twice as big while they were still in the nest.

In the other nesting grounds, we also found some babies as small as 14 inches and some as big as three and a half feet long—but none smaller than 14 inches and none longer than three and a half feet. Thus we concluded that the babies were about 14 inches long when they hatched. And it seemed that they stayed in the nest until they were about three and a half feet long, until they had more than doubled in size. Now, it takes a bird a month or two to double in size. But it takes an alligator, a fairly sophisticated reptile (alligator mothers do care for their young to some extent), a year. No known creature stays in a nest for a year. So if these animals did stay in the nest, as we concluded, and if they grew at the rate of coldblooded animals, there was an insoluble problem.

I can't imagine that the babies left the nest. For one thing the nests that didn't have fossil skeletons, the nests in which babies had successfully hatched and grown, yielded crushed eggshell, not intact eggs. In other situations, such as the *Protoceratops* find in Mongolia, the eggs were not smashed. Those nests yielded half-eggs, broken in hatching and then left undisturbed. Dinosaurs were probably as diverse in their habits as they were in size and appearance, and the protoceratopsians must have left the nest as soon as they hatched. Otherwise, in walking and moving around in the nest, they would have trampled their eggshells. At the nesting ground of maiasaurs, the eggs were pounded into fragments—pounded, I'm sure, by the feet of baby dinosaurs.

I might point out also that the maiasaurs were not alone in their world, or in their nesting ground. There were lizards, whose bones we have found in various parts of the anticline. Similar lizards today eat eggs. I'm sure they ate eggs then. In fact, almost anything except a strict herbivore will eat an egg if it gets the chance. There were also small, wolf-size predatory dinosaurs, whose fossils we have found in other areas of the anticline. I suspect they raided the nests in groups, trying to avoid the adult maiasaurs and snatch a baby or two. The inevitable presence of these predators is another reason to believe the young stayed in the nest. Baby alligators get eaten in great numbers, and the young maiasaurs were nowhere near as toothy and nasty as they are. A bunch of baby maiasaurs walking around alone would have been like meals-on-wheels for the carnivores. They wouldn't have to invade a nesting ground; they could just sit and wait for the *plat du jour* to waddle by.

And what if they weren't alone? What if they were waddling along with the adults? Well, in that case, creatures a foot and a half long would have been around other creatures whose feet were a foot and a half long. The adults could have weighed three tons. The babies weighed three or four pounds. As I pointed out in Chapter 2, they would have had a hell of a time not getting trampled. With family like that, who needs predators?

No. They had to stay in the nest. And somebody—most likely Mama, but conceivably Papa—had to bring them food. But a year is much too long for any parent to be running back and forth from the berry bushes, so the maiasaurs had to grow fast, like birds, and be out of the nest in a month or two. Then the whole setup would make sense. And the best way to grow like a bird is to be warmblooded like a bird. I wasn't completely convinced by this evidence that all dinosaurs were warmblooded, but for the maiasaurs, at least, I was having trouble coming up with an alternative.

The fossil evidence gave me not only these scientific conclusions but also images of the individual dinosaurs and, in fact, of a whole social scene. I could see it, in living color, as I sat on one of the anticline's hills amid the grass and the gophers and the prickly pear.

PICTURE YOURSELF ON A PLAIN, flat like the Great Plains, but a coastal plain. The sea is 100 miles away. Around you are numerous small streams bordered by dogwoods. There are vast flat expanses of something like raspberry thickets. In among the nests the ground is beaten down from the tromping of large dinosaurs. These are, let's say, reddish brown, with pale undersides. They are as long as elephants, but much thinner in body, and they move more fluidly, bobbing their heads the way birds do. From a distance, you can see these large dinosaurs moving in and out of the trampled area. Sometimes you can hear them, bleating perhaps, or honking. If you move closer, you see that they are eating berries by the stream. At the nests they open their jaws wide and regurgitate the berries.

In the nests are newly hatched, foot-and-a-half-long, uncoordinated, squeaking baby reptiles. Their faces are flat. Their eyes are big. They are all crowded together and very noisy. They scramble over each other for the food their mothers bring them. Sometimes the adult dinosaurs simply sit or lie by the nests. They also spend some time eating the leaves of the dogwoods and branches of the evergreens for themselves. The dinosaurs show the kind of alertness and playfulness you would expect from lions with their young, or from wolves, or horses, because like those animals they have the high metabolism of warmblooded creatures. Any feelings they might have, however, any primitive emotions, are impenetrable. To look into their eyes is to look into a lizard's eyes, or a bird's eyes, not the seemingly revealing eyes of the family dog.

The nests are individual mounds of mud built by the adults with their forelimbs and hind feet and then scooped out. Over one nest, in which the eggs have not hatched, is a layer of rotting vegetation. The dinosaur that laid these eggs does not sit on them or even near them. She wanders from the nesting ground to the stream and back again. She walks on her hind legs but drops down on all fours from time to time to investigate the nest. When she lies down, she may sleep or doze in some fashion; perhaps she is simply still, inactive but attentive, like a bird at rest.

That much, I think, is easy enough to imagine. But at some point later that year, perhaps when the babies were grown enough to leave the nests, I imagine something else. A herd of maiasaurs would pass by, thousands and thousands of them. The noise would be tremendous. But even more impressive would be the sight of all those heads bobbing, ten thousand ducks' bills gliding forward and back like the heads of pigeons, making the whole huge herd ripple in the sun like a mirage.

It is only in my imagination that I see it, of course, but the herd is no mirage. It was there, and we found the bones to prove it.

CHAPTER 5

THE HERD

Only in our first year did we camp by the Teton River and drive to the Peebles ranch. In 1980, courtesy of the Peebles family, we moved to a spot right on their ranch, smack dab in the middle of the anticline. It was also in 1980 that the National Science Foundation took over the funding from Princeton and our grants began to grow steadily. More money meant a larger crew, more finds and more evidence of how rich the site was, thus better grant applications and still more money.

More money also meant construction. Bob, who had a passion for carpentry, was the builder. First he built a kitchen, a frame on which to put a tarp so that we could cook and eat our meals out of the sun and rain. Then he built a root cellar, to supplement our gas-fired refrigerators, and the coolers. Along the way, he built big wooden boxes to store food away from the ground squirrels and a loading platform for the huge, plaster-covered chunks of rock and fossil we brought out of the field and transferred from one truck to another. In the later years of the camp, he built a framework to hold a truck inner tube full of water so we could have warm showers through solar heating. This meant we were no longer limited to one shower a week as we had been before; we got that one at the campground in Choteau ($2 per shower) on our usual Friday trip to town for groceries, telephone calls and ice cream cones.

I found camp very comfortable, except when it rained; then the ground, which was mostly bentonite, turned to a heavy, sticky substance something like wet cement that clung to our boots and made us walk as if earth gravity had just gone up a notch. Rain also reminded us how isolated you can be in Montana, even when you're on somebody's ranch. Choteau was about 12 miles away on a good gravel road, but getting to that road meant driving on well-worn ruts through cattle range for about a quarter of a mile. This was fine in dry weather. In rain the ruts turned to muck, as did the rest of the range, but they were still the safest place to drive because they had been packed down. The surrounding ground was softer and even more likely to bog down a truck or car. We once had visitors who had brought along a one-year-old and ran out of diapers in the middle of a day and a half of hard rain. I drove one of them out in their rent-a-car, which, fortunately, was front-wheel drive, with my foot on the accelerator and the rear end fishtailing for the whole quarter-mile. If I had stopped, the car would have sunk in the mud. Nor would one of the four-wheel drive trucks or jeeps have helped me. In that kind of mud, all you get with four-wheel drive is four wheels digging down deeper, faster. You end up high-centered, with the drive train sitting on the mucky center of the road and four wheels spinning free.

Naturally, with each success at the anticline, my position at Princeton improved. I became less and less a preparator, except of my own fossils, and more and more a principal investigator (as the National Science Foundation terms it) in charge of a research project. Vertebrate paleontology was not a high priority for Princeton University, however; there was a small museum of natural history that was part of the geology department, and there was only one professor, Don Baird, with one or two assistants. In 1982, partly because of the limited future at Princeton and partly because I wanted to go back home, I moved to Montana State University, where I became curator of vertebrate paleontology at the university's Museum of the Rockies and an adjunct professor. The museum was flourishing, and it was particularly interested in supporting paleontological research. Here, in Bozeman, I knew I would be close to my field research. I even managed to get an advanced degree once I moved back. In 1986 the University of Montana, my one-time intended alma mater, gave me an honorary doctorate. That same year Princeton University shut down vertebrate paleontology, except for Don Baird, who stayed on until his retirement in the spring of 1988 and who is now a research associate with the Carnegie Museum in Pittsburgh. Princeton closed its small museum of natural history and gave its entire fossil collection to Yale, including the finds from the Willow Creek anticline. Yale has been kind enough to let me keep those fossils on loan in the research collection at the Museum of the Rockies.

One of the major features of that collection is a group of strangely battered adult maiasaur bones from the anticline. I have thousands of these bones, cataloged and arranged on floor-to-ceiling metal shelving. They all show the same pattern of breakage and wear; they're all the same gray-black color, which indicates similar conditions of fossilization; and, of course, they all came from the same site. They belong together, although this fact was not immediately apparent to us when we started finding them. Only in retrospect, after analysis of the finds, could we see the whole site clearly.

You might say that the course of the dig at the Willow Creek anticline was like the course of one of the braided streams that, 80 million years ago, in the Cretaceous, deposited the mud on the dinosaur bones to preserve them for us. Those streams consisted of several individual channels, really separate streams themselves, that wove back and forth between set banks, crossing each other, joining and separating. That's what work at the anticline was like—separate streams of investigation weaving in and out of each other, joining and separating.

IN 1979 THE BRANDVOLDS were still prospecting on the Peebles ranch, looking for a big, relatively intact dinosaur skeleton to reconstruct and put on display in their rock shop. And in July they thought they had found one. They had uncovered a big femur and an equally big humerus on a ridge about a half-mile from the part of the anticline where Bob and I and the crew were searching for maiasaur nests. The task of exhuming a full dinosaur skeleton in good enough condition to reconstruct was a major one, however, and they needed our help. In return, they agreed to map the site and keep track of each individual bone fragment so we would know what had come out of that deposit. They would have the bones, and we would have the information.

Before they even got started on any serious digging, I put my crew on the site for a week to take off some of the overburden. (In plain language overburden is dirt, the soil that lies on top of rock in which the best-preserved bones are likely to be embedded.) In that week, the crew found 65 bone fragments. These fossils were poking out of the rock, directly under the layer of dirt. The Brandvolds worked that site through the summer of 1979 and the following fall and winter, carefully mapping and photographing the site as we had requested. At the beginning of our 1980 field season, I looked over what they had found and saw that they were in for a disappointment. There were the remains of at least five individual dinosaurs, three of which were juveniles. That is to say, they were longer than 4 feet, which is as big as the babies got in the nest, but were less than 10 feet from tip to tail, making them less than half grown. The Brandvolds had uncovered a rich but messy deposit. There were probably well over a hundred individual fragments—pieces of crushed, distorted and badly broken bones. The odds of getting a whole composite dinosaur out of the site were very small, even if you worked on it for several years, but for us the dig was worth something because here were the remains of a group of adult and juvenile dinosaurs.

We could tell from the bones that these were hadrosaurs, and we hoped they might be maiasaurs and that we might have found not just a chance collection of bones but some kind of social grouping. Perhaps, we thought, these dinosaurs nested in colonies and lived in small family groupings. It was just speculation, but it's this kind of speculation that fuels the imagination and gives you the energy to dig, and dig, and dig, until you find out what you've actually got. The Brandvolds turned the site over to us to excavate as part of the dig, and from that point on it was our crew who did all the prospecting and digging at the anticline. As it turned out, we never got enough for even half a dinosaur from that particular spot, but then that wasn't what we wanted.

We wanted to know if the dinosaurs preserved, albeit badly preserved, in this spot had been together in life. If this was really a family unit at the Brandvold site, then all these dinosaurs had died together and been buried together. We had to think if that was possible, and if so how. We also had to consider the possibility that perhaps this was a random collection of bones of the sort that might accumulate in the bend of a river where sediment and detritus were deposited by currents. By the end of 1980 we had 200 bones from that site, representing at least eight individuals, all in very bad condition, embedded in mudstone with a lot of volcanic ash. I knew the dinosaurs were all hadrosaurs, and all the same kind of hadrosaur, but I couldn't make a more precise definition. There was certainly no indication that these bones had been brought together by accident. They were not in a river or stream bed. In fact, it wasn't clear what kind of deposit they were in. It was mudstone, all right, but we couldn't tell where it came from. John Lorenz, who was working on the dig in 1980, thought the animals might have been caught as a group in some kind of catastrophic mud flow—a flood, but of mud instead of plain water. Lorenz was a Ph.D. student at Princeton at the time, and he was studying the stratigraphy of the Two Medicine formation. (Stratigraphy is basically the study of the various layers of rock, and what has happened to them. A stratigrapher sets you straight about which beds are oldest, for instance. It's part of figuring out the geology of your site so you have a framework to fit the fossils into.) Lorenz' theory could help explain some odd facts about the site. The bones were in awful condition. Some even looked as if they had been sheared lengthwise. However, right next to a badly damaged bone would be one that was untouched. Furthermore, we found some of the bones standing upright. Bones caught in water or lying on the ground and buried by sediment don't stand up vertically. But if the creatures had been caught in a mud slide, they might have just been bashed to pieces and left in these odd positions as the mud settled.

Lorenz suggested that the mud slide could have been caused when a volcano erupted, spewing out ash. The ash found in the Brandvold site might have clogged a lake, turning it into thick, viscous mud. Had the lake breached some natural barrier, the resulting flood would have produced this kind of mud slide. We knew that there had been large lakes in the area at earlier time periods, and there could have been one at this time as well.

This was a speculative explanation, and both Lorenz and I recognized that it raised as many problems as it solved. In the Brandvold site, for example, there were hardly any small bones, such as toes and fingers. And there was only one skull, which we found much later. More of these kinds of bones would presumably have been there if the animals had been caught and buried alive. Furthermore, the damage was not really of the sort that could happen to living animals. How could any mud slide, no matter how catastrophic, have the force to take a two- or three-ton animal that had just died and smash it around so much that its femur—still embedded in the flesh of its thigh—split lengthwise? We left the problem unsolved, because we didn't have much choice. And, as happens with such problems, it got bigger.

ONE DAY IN 1981 MY SON JASON, who was eight years old at the time, took a small hike while the rest of us were working on the Brandvold site. Jason has made a number of paleontological and biological discoveries, including one I remember quite well that occurred when he was four and we were in the Judith River formation. He was some distance away, and he informed me that he had found a cute baby lobster. I told him not to touch it and hurried over to him. He just let the animal crawl around in the rock fort he had built around it until I got there. I saw exactly what I expected to see, the closest thing to a lobster that the badlands have to offer: a scorpion. I explained all about scorpions to Jason, and he took the information in stride.

He kept on exploring and discovering things—horned toads, Indian arrowheads and dinosaur fossils. On the walk he took in 1981, he found nice, big hadrosaur fossils. We named the site Nose Cone, for the simple reason that Jason, who found it, thought the rock there looked like the nose cone of a rocket. On my digs, whoever finds a site, names it. When we got around to excavating the Nose Cone site, we found adult and juvenile bones of the same kind found at the Brandvold site, and in the same condition and state of disarray. Even more intriguing, it was obvious that these two bone deposits were connected. You could walk from the Brandvold site to Nose Cone and follow exposed bones all the way. Both sites were on the side of a hill, and we poked into the hill here and there to see if there were more bones. There were. The two sites were clearly on the same fossil horizon. So it looked as if we had a bone bed that extended for at least an eighth of a mile. The family of hadrosaurs was growing.

We found Camposaur in 1981, too. Or Camposaur found us. It more or less stuck itself right in Wayne Cancro's back. One of our first steady volunteers, Wayne joined us in 1981 when we were in our second year of camping on the anticline itself. He arrived early and picked a choice spot for his tent, planned how to dig his trench to prevent accumulation of rainwater around his head, and proceeded to attempt to hammer in his stakes. He had a lot of trouble getting them into the ground because he kept running into bones. In itself, this wasn't such a big surprise—we were finding the odd bone fragment around camp all the time, and there were two fairly large bones on a little rise in camp that we were trying to get out intact. So Wayne kept working to get his tent set up.

When he finally did get his stakes in and went inside to lie down after the effort, he suffered the Princess and the Pea phenomenon. There were bones poking him everywhere. It was Wayne's extreme discomfort, which led him to move his tent, that led us to start digging seriously in camp. We figured if it was that bad, it had to be good. So we began to dig, and eventually we had a pit in the middle of camp about 20 feet by 30 feet and in places 3 feet deep. From that pit, between 1981 and 1984, we pulled out 4,500 maiasaur bone fragments. They were all like the bones from the Brandvold and Nose Cone sites—black, rock-hard but crumbly and often fragmented. That total represented about 30 individuals. The bones were of adults and half-grown juveniles, the same kinds of animals found at the Brandvold site. Wayne called this site Camposaur on the basis of its location—right in the middle of camp.

Like all our other pits of any size, Camposaur was divided into a grid of squares about three feet on a side, with each square mapped to show the fossils as they lay in their original position. The actual digging proceeded in typical fashion. First we shoveled dirt, carefully, until we reached the level of the preserved fossils. Then, gradually, we worked on the pit. One or two people would take a small area and brush and

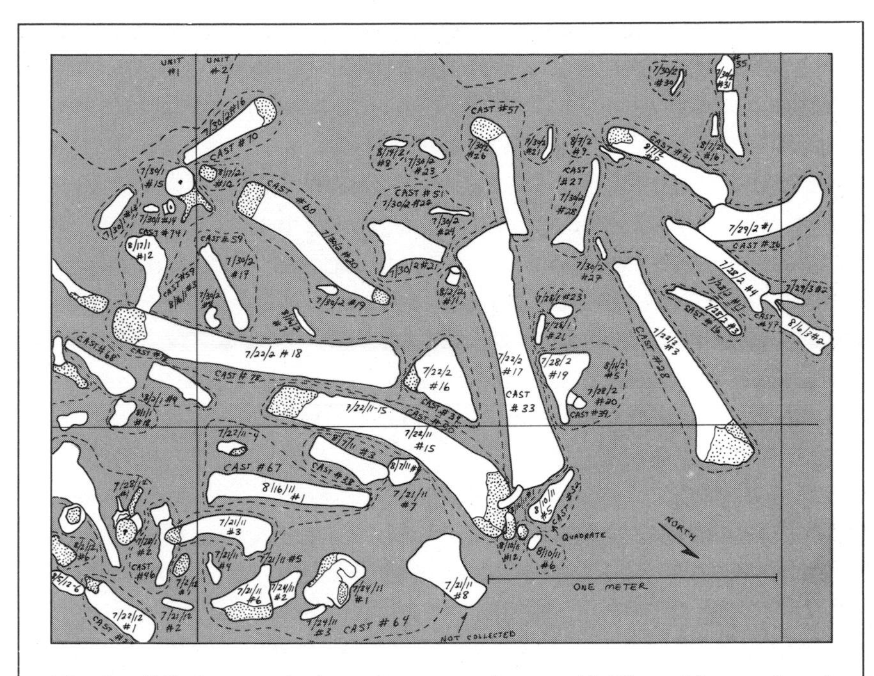

The fossil finds at each site were mapped on a grid. The grid reproduced here, of the Brandvold site, reflects the profusion of bones common to Camposaur and other parts of the big bone bed.

sweep dirt away with whisk brooms if a bone was already exposed, or gently poke and pry with an ice pick, carefully, if no fossils were sticking up from the ground.

When we found a fossil, a legbone, for instance, we used the ice pick, toothbrush and whisk broom to loosen the dirt on and around it. Then, as each section of the bone was exposed, we painted it with polyvinyl acetate to help hold it together. Once the top of the bone was exposed, we dug and poked around it with the ice pick, all the while cleaning and painting each section that we exposed. Eventually we would have the bone resting on a pedestal of dirt or rock, at which point we put a plaster cast on it.

We made the casts the old-fashioned way, by soaking burlap strips in wet plaster of Paris and wrapping them around the bone. When a cast had hardened, we took out the bone by breaking through the pedestal. What we ended up with in this process was a whole bone protected by a cast. Smaller fossils found close together would all be included in one cast. Some of the better-preserved nests were wrapped in plaster, with all the eggs and other fossil bones in one cast. In the winter, back in the laboratory, we would take off the casts and begin the preparation of the bone—cleaning it again, identifying it, putting it back together if it was broken and coating it with varnish to help preserve it. Then the bone would be cataloged and saved for study either then or later.

BY THE END OF THE SUMMER of 1981, we had a lot of similar bones to think about. The Brandvold site and Nose Cone were clearly part of one bone bed. They were in the middle of the anticline on the north side of the eroded center. If you crossed over the center, you came to camp. There you found Camposaur. And if you continued on, you found the children's dig, which we had created near the top of a steep ridge where Bob Makela liked to demonstrate how well his Toyota jeep handled inclines. He'd get a passenger, put the Yoda, as it was called, in four-wheel drive, and then shoot up the hill until he brought it to rest at a 45-degree angle. Kids loved this ride. We created the children's dig not because we wanted to employ child labor but because Jason was there for each field season, Bob's son Jay came down sometimes, and we had a lot of visitors who brought their children. Often they would stay for only a few days, but we needed a place for children to work, where they could actually do some good. The children's dig was a pit on the same horizon as Camposaur and so similar in the bones that came out of it that we could afford to take some chances with a bone or two getting smashed or lost. We did keep an eye on it and mapped the bones in their original locations before the kids started on them with ice picks and whisk brooms to liberate them from the mudstone.

It was fairly obvious that the Brandvold site and Nose Cone were part of the same deposit. And it was equally clear that Camposaur and the children's dig were part of one deposit. But was it all the *same* deposit? Late in 1981, we had our answer.

I had been sitting on the hill behind the kitchen one day, looking down on Camposaur and north toward the Brandvold site, when the thought came to me again, but this time more forcefully, that each of these deposits had the same black, battered bones of adult and juvenile maiasaurs.

I decided to try a simple test right then. What I used was a Jacob staff—a five-foot-long board with a Brunton compass attached. The staff is used to measure vertical distances between beds of rock. The Brunton compass, a common gadget in geological fieldwork, has a level and can be set to compensate for a slope in the ground so that you get a true reading of vertical distance. I started my measurements at the bottom of the anticline, in the first hadrosaur nesting ground. This nesting ground was the lowest fossil layer we identified in the anticline, and it clearly existed on both sides of the eroded scoop that separated the Brandvold/Nose Cone bones from the Camposaur/Children's bones. I measured from the first nesting ground to the other deposits on each side. And I found that the vertical distance was the same to the Brandvold/Nose Cone site as it was to the Camposaur/Children's site. By this fairly crude test (the Jacob staff is an instrument for quick estimates, not precise information), I confirmed my suspicions. All four sites were on the same fossil horizon.

We didn't get a final confirmation for this conclusion until the end of the summer season three years later in 1984, our last season at the dig. The season's work was just about over and I was walking around the anticline with Will Gavin, the graduate student who had done a study of the geology of the Willow Creek anticline. We were up on the ridge, where the children's dig had been, and he was showing me some of the peculiar geological features that he had found. In the hillside just above the bone deposit that the children had been working on, Will noticed an ash bed. This wasn't something he had found previously. He saw it as we were standing there. Ash itself was not unusual; we had found it at the Brandvold site. But this was a definite layer of ash sitting just above the bones, something we hadn't noticed at the other sites.

We immediately set out to check them. The first stop was camp, where we realized that we were standing on the ash layer. Bentonite, the stuff that turns to wet cement in the rain, is a mineral that is, in essence, volcanic ash. At the Camposaur pit we could see the layer nicely delineated, just as it had been at the children's dig. We spent all that day checking the other sites, and we found the same ash layer precisely the same distance—18 inches—above the layer of bones. This held for the Brandvold site, for Nose Cone, for two other pits of bones, which, up until that point we had not connected to the big deposit, and for some test pits we had been digging to see how far the deposit extended.

There was no question anymore. We had one huge bed of maiasaur bones—and nothing but maiasaur bones—stretching a mile and a quarter east to west and a quarter-mile north to south. Judging from the concentration of bones in various pits, there were up to 30 million fossil fragments in that area. At a conservative estimate, we had discovered the tomb of 10,000 dinosaurs.

I should point out that, although we suspected from the start that these were maiasaur fossils and we knew they were hadrosaurs, it was when we found parts of several skulls in Camposaur, in 1982, that we positively identified the animals buried in this bone bed as maiasaurs. At that time, in the early '80s, there was no other single deposit known with so many fossils of one kind of dinosaur. And it was just one kind. In all the years and all the pits we dug in that big bone bed, the only other things we found were carnosaurs' teeth and one small dinosaur of unknown variety that was rolled up in a fossilized mudball.

What could such a deposit represent? None of the bones we found had been chewed by predators. But most of the bones were in poor condition. They were either broken or damaged some other way, some broken in half, some apparently sheared lengthwise. They were all oriented from east to west, which was the long dimension of the deposit. Smaller bones, like hand and toe bones, skull elements, small ribs and neural arches of vertebrae, were rare in most of the deposit. At the easternmost edge of the deposit, however, these bones were the most common elements. All the bones were from individuals ranging from 9 feet long to 23 feet long. There wasn't one baby in the whole deposit. The bone bed was, without question, an extraordinary puzzle. First there was the terrible condition of the bones. As early as the first Brandvold site, we thought that a mud flow might have done this. However, on reflection, the condition of the bones argued for something other than animals just being buried alive, even in a vicious mud flow from a breached lake. As I mentioned before, it didn't make sense that even the most powerful flow of mud could break bones lengthwise when they were still padded in flesh and tied together by ligaments. Nor did it make sense that a herd of living animals buried in mud would end up with all their skeletons disarticulated, their bones almost all pointing in one direction and most of the small bones at one edge of the deposit. It seemed that there had to be a twofold event, the dinosaurs dying in one incident and the bones being swept away in another.

Jeff Hooker, an MSU graduate student, had worked on the fossils from the big bone bed and was the first to question seriously the idea of a herd dying in a mud flow. He was studying the bones from Camposaur—4,500 bones, representing 27 individual dinosaurs—and he began to notice certain things about the damage they had suffered.

First of all, the ones that had broken showed clean breaks, not jagged, splintery breaks. Fossil bones break this way, cleanly, like rocks. Fresh, or dry but unfossilized bones splinter. The Camposaur bones looked like they had been broken after they had been fossilized.

Furthermore, Hooker thought the bones that appeared to have been sheared lengthwise had not been broken at all. He suggested that the bones had lain on the ground, as they would have if the dinosaurs had died and rotted aboveground, and that because this was a volcanic environment, which the presence of the volcanic ash suggested and which is consistent with the known geology of the area, the groundwater would have been very acid. That groundwater could well have eaten away or dissolved parts of these bones, leaving them looking as if they had been neatly sliced lengthwise. Perhaps before or perhaps after the acid had partly dissolved these bones, fossilization had begun. Fossilization can occur before burial. That same groundwater could have been rich in minerals, starting the fossilization when the bones absorbed these minerals. It was after fossilization that the bones were swept along, in something like a mud flow, and deposited in their current location.

The layer of volcanic ash resting just a foot and a half above the bones was the key to how all these events could have occurred. You may remember the devastation and widespread ashfall caused by Mount Saint Helens. That was a little volcano. Volcanoes like that were a dime a dozen in the Rockies back in the late Cretaceous. There were much bigger volcanoes, in the Rockies, to the west of the site, and also south of the site, in what are now the Elk Horn Mountains near Great Falls. What Hooker suggested was that the dinosaurs, a herd of *Maiasaura*, were killed by the gases, smoke and ash of a volcanic eruption. And, if a huge eruption killed them all at once, then it might have also killed everything else around. That would explain the lack of evidence of scavengers or predators gnawing on the bones. They would have been dead, too, or perhaps they would have fled the heat, gases and fires of the volcanic eruption, not returning until the corpses had rotted. Without carnivores what we would have seen from a helicopter would have been a huge killing field, with the rotting corpses of 10,000 dinosaurs. The smell would have been overpowering. And the flies would have been there in the millions (there were flies in the Cretaceous). It must have been one hell of a mess.

Over time, of course, the stench disappeared and the killing field turned into a boneyard. Perhaps beetles were there to clean the bones. The bones lay in the ash and dirt. Some fossilization occurred, as well as some acid destruction of the bones. Then there was a flood.

This was no ordinary spring flood from one of the streams in the area, but a catastrophic inundation. Perhaps, as John Lorenz thought, a lake was breached, turning the field of death—now covered with partially fossilized, partially dissolved skeletons, unconnected by ligaments, flesh and skin—into a huge slurry as the water floated the bones, mud and volcanic ash into churning fossil soup. The bones of the maiasaurs would have been carried to a new location and left there as the floodwaters or mud settled. Had this occurred, the bones would have acquired their uniform orientation, and the smallest pieces, weighing the least, would have been carried the farthest. Finally the ash, being light, would have risen to the top in this slurry, as it settled, just as the bones sank to the bottom. And over this vast collection of buried, fossilized dinosaur bones would have been left what we now find—a thin but unmistakable layer of volcanic ash.

That's our best explanation. It seems to make the most sense, and on the basis of it we believe that this was a living, breathing group of dinosaurs destroyed in one catastrophic moment. The destruction is, of course, astonishing. But to be amazed by that is really to skip over something much more startling, and that is the herd of maiasaurs.

The notion of a herd of dinosaurs is not a new one. The paleontologist Roland T. Bird suggested that the sauropods might have been herd animals.[1] There are footprints that suggest dinosaurs moved in groups, and there are numerous cases of a group of fossils of one sort of dinosaur found together.[2] But there was nothing of this order—no pile, in one spot, of 10,000 dinosaurs.

The question is, was this just a bunch of maiasaurs eating together or did they have some social structure? Did they stay in this kind of herd all the time? What was the relation of the herd to the nesting ground? In other words, what do all these bones say about how these dinosaurs lived their lives? I can't say I have the answers to these questions. The physical data we can be absolutely certain of. Some forms of behavior, like colonial nesting and parental care of the young, are as certain as they can be for 80-million-year-old animals. When we get to the social structure of the maiasaur herd, we are suddenly on much shakier ground. I have clues, and I can offer them, but speculations and guesses are all I can build on them.

Jeff Hooker did some preliminary sorting of the bones from the Camposaur pit. On the basis of his work, it seems that the dinosaurs in the herd ranged from 13 feet long to 24 feet long. This means the youngest dinosaurs are missing. We know that the young grew to 4 feet long in the nest. A reasonable guess is that they reached something like 8 or 9 feet in a year, and we have found fossil remnants of 9-foot-long maiasaurs. Why didn't we find them in the herd?

Perhaps the babies of that year were still in the egg or in nests when the volcano erupted, or perhaps nesting had not even begun. This bone bed is not on the same layer as any of the hadrosaur nesting grounds, so we have no hope of finding those nests, eggs or babies. However, even if we were to assume this catastrophe caught a herd at nesting time, that still would not account for the previous year's young, animals who would have been nine feet long at the time. Where were they?

One answer is that they may actually have been in the herd. There is conflicting evidence on this point. Camposaur, the site that Hooker looked at, contains none of these nine-footers; however, at three other pits, we have found nine-footers. One pit, called Fire Ridge, is a mile away from the end of the big bone bed but may still be part of it. It has the telltale ash bed. (So the herd may have been much bigger than we realize.) The other two pits, Sacred Slump and Worthless Wash (I said we had weird names for our sites), both are clearly part of the bone bed. Fire Ridge yielded an adult and four nine-footers, and the other two each yielded an adult and three nine-footers. Now, Sacred Slump and Worthless Wash are on the edges of the big bone bed. And invariably the pits dug along the edges of this deposit show bones that are better preserved than those in the middle. It may be that the mud flow was something like a professional football game, and that the middle of it was like the point where the defensive and offensive lines meet and bones (human or dinosaur) are shattered. Small bones wouldn't have a chance. On what we might call the sidelines, however, the smaller bones would be less subject to breakage because the flow might be less forceful or because the concentration of bones would be lower. Or perhaps, because of vagaries of current in this slurry, some smaller bones might migrate to the sides of the flow.

That's a guess. Another guess, an even bigger one, is that the nine-footers stayed with their mothers in small family groups and some were joining the big herd when the volcano struck. And as long as we're involved in speculation right now, I should point out that nobody knows for sure that these dinosaurs would have produced young each year.

I hope in future field seasons to explore the edges of the big bone bed more thoroughly and get a bigger sample of bones. Until then, we can only talk about what we know. So far the story of maiasaur growth and social life seems to be hatching and growth in the nest, followed by a period of mystery, followed by joining a large herd or at least an aggregation. In biology a "herd" has a clear social structure; it's not just a random aggregation. In some species the males form the

Like the American bison millions of years later, herds of thousands of maiasaurs

traveled the late Cretaceous plains in search of forage.

perimeter of the herd and the females and young stay in the center. We have no evidence that this was true for the maiasaurs, because we don't know how to tell a male from a female, for one thing, and because the collection of fossils tells us nothing about how the dinosaurs arranged themselves in life. They were together, I'm sure of that, but I have no idea who had what place in line.

How did they behave in their aggregation, then? How did they mate? How did they defend themselves? Because these were social animals, and herbivores, I suspect that some of their behavior might have been parallel, in at least rudimentary form, to what we see in social herbivores today—namely, competition among males for female mates. Perhaps the problems that large herbivores must solve in terms of finding food, reproducing, and defending against predators are the same for any large herbivore and call forth similar solutions. We know at least that the maiasaurs had these huge aggregations, throwing all the males and females together and opening up the possibility for sexual competition. We also know that some duckbills had bodily characteristics such as frills of skin and cartilage running down the spine. Others, the lambeosaurs, had the elaborate skeletal crests. Why?

The answer I hold to is that the purpose of these characteristics was to attract mates.[3] And I would say the same for the horns of *Triceratops* and its kin and the clubs of the ankylosaurs. I believe sexual attraction, not defense, was the reason these characteristics evolved. This is, in fact, one of my pet peeves. To me there is no subject on which so much nonsense has been heard as defense among dinosaurs. It's not just the paintings of *Triceratops* battling *Tyrannosaurus* made for popular consumption. Scientists, too, assume that horns and clubs were weapons. This is seldom the case among animals today, and I doubt things were different in the Cretaceous. All the existing herbivores with great antlers developed them to attract mates and conduct sexual combat between males. Occasionally, very occasionally, the horns may be used, opportunistically, to fend off a predator, but I doubt that this usually succeeds. An elk is better off using its feet rather than its horns to attack wolves. Bighorn sheep use their magnificent horns to butt each other. Elk and moose do the same. And so, I bet, did *Triceratops*.

I see no sign of combat among the maiasaurs, but I can imagine females making nesting sites and waiting there for dominant males to come to them. Who knows how the males might have established dominance—by vocalization, perhaps, or some kind of display of their frills, to seem larger. As for defense, I think the hadrosaurs and the lambeosaurs, all of them, and certainly the maiasaurs solved that problem by gathering in the herds that we now know some of them formed. This is the common defense in large mammalian herbivores today. The dinosaurs were also large terrestrial herbivores, and they may have come up with this solution first. Of course, predators still took their share. It doesn't pay to think of evolutionary adaptations for defense as if the animals themselves were deciding how to fend off those nasty meat-eaters. The process would go more like this:

Imagine early herbivorous dinosaurs with no innate instinct to gather together. Predators chase and kill whichever ones they can, finding them alone among the berry bushes. Suppose that some of these dinosaurs are born, through random shuffling of the genes, with the desire to stay in groups. These social dinosaurs do better in the struggle for survival, perhaps because it seems easier to a predator to attack a lone animal, or because in a group there are a number of eyes and ears to warn of attack. Consequently, the social dinosaurs produce more offspring, all of whom share the genetically programmed instinct to stick together. Over the course of time, the loners disappear. Predators adapt to following the herd, picking off the weak, the old, and the young who stray—in effect culling out the poorer specimens. The result is a kind of balance that is achieved, a state of equilibrium, and the word "defense" may be somewhat misleading.

I wonder a bit about how these kinds of herds affected the environment. Certainly the herds had to keep on the move. They must have stripped one area and then moved on to the next. I don't think *Maiasaura* migrated down to the sea and back, since her fossils have never been found in lowland areas. I suspect, although I have no positive evidence for this, that the maiasaur herds migrated seasonally on a north/south pathway. Perhaps it's hard to imagine dinosaurs in great herds surviving on the sedges and berry bushes that were prevalent in the upper coastal plains during the late Cretaceous. But think of the bison. One estimate has it that in North America, at the start of the nineteenth century, there were 60 million of them. All they ate was grass.

Given the fossils we have, I think that when all the scientific papers are written, when the studies of the bones are completed, *Maiasaura* may end up being the best-documented dinosaur paleontology has seen. Because of the wealth of bones of this one species we have the chance to learn more about her, from basic morphological structure, to evolution, to social behavior, than we know about any other dinosaur. This wealth of information that she has presented us with is a remarkable yield for one dig. But she is not the only dinosaur to come out of the Willow Creek anticline. The dig produced fossils of another dinosaur as well. And I don't mean just a few bones. I mean eggs, skeletons and nesting grounds—another entire dinosaur world.

CHAPTER 6

EGG MOUNTAIN

Off the west slope of the anticline rises another hill—obscured on the surface, like the anticline itself, by geologically temporary rises, ridges and arroyos. We have only one slope of this hill left, however, so it does not qualify as an anticline. It's called a monocline. This feature is in the same two square miles in which we made all our finds on the Peebles ranch, and it dates from the same time period, so we usually described the whole dig as the Willow Creek anticline.

On the surface, covering the monocline, is a small, round-topped hill. We found this hill in 1979 during the first field season, the same season in which we began to uncover the extent of the first maiasaur nesting ground and in which we found the first clues to the big bone bed of maiasaurs. The entry in my field journal for July 11 reads: "7/11/79—Weather." In other words, it rained that day. We spent the time in our tents and our cars (when it comes right down to it, cars are drier than either tents or teepees), bored, cold and damp. The next day the sun returned and along with it the heat that we had been experiencing. It was intense enough to make the cattle range wobble in the haze. The entry for that day begins: "7/12/79—Back to Worthless Wash and Sacred Slump."

We were camped then on A. B. Guthrie's land near the Teton River, so we drove to the ranch and parked our cars by the side of the gravel road next to the Peebles' land. Then we walked to Worthless Wash and Sacred Slump, the two pits that later turned out to be part of the big bone bed. At the time they were just deposits of bones to us. We were working them, the Brandvolds were working the "Brandvold site" (we had just gotten the overburden off for them), and we had just spent several weeks searching for, and finding, maiasaur nest sites. Paleontology is a bit like war in that each day you marshal your forces and send them where they seem to be needed at the time.

Worthless Wash and Sacred Slump were both on the anticline proper. On our way to the anticline, each day we went around—or over, if we were energetic—a small hill that stood in our way. That day, by the time the afternoon arrived, we had taken out of Worthless Wash and Sacred Slump a pubis, a tibia, part of a jaw and a sacrum, and we had also uncovered a cervical vertebra. All of these were maiasaur bones. While we were poking around with our ice picks and brushes, two human figures materialized in the distance, shimmering on the horizon as if Scotty had just beamed them down. You don't often see people strolling through the cattle range. A truck or four-wheel drive vehicle, or more traditionally a horse, is the usual means of transportation. We were a mile or so from the road, and we watched the two approach. Periodically they would stop, take measurements and put a flag in the ground. Then they would walk on, stop, measure, and plant another flag. By the time they reached us, we could see they were not apparitions but rather a man and a woman carrying surveying equipment.

They worked for a seismic crew that was doing oil exploration for Shell. Their job was to lay out a line for the seismic crew to follow.

Once they had set this line, other men would come along with big shaker trucks or explosives. They would follow the line, and at each flag set off an explosion or stop the shaker truck, have it let down what looked like huge pile drivers on its sides and shake the earth. I don't mean just jiggle it; the whole point of the exercise is to mimic an earthquake and then study the seismic waves that it generates. These waves travel differently in different types of rock, so that a certain kind of formation will have a seismic signature. What oil companies look for is a signature that says petroleum. They like to find spongy sandstone, perhaps right beneath some shale. This grouping of rock might represent the remnants of a beach and heavily vegetated lagoon. The partially decayed plants and animals from the lagoon may have been transformed into crude oil, which can be absorbed and held in spongy stone the way water is held in an aquifer. So one way to look for oil-bearing rock is to draw a straight line across miles and miles and then go along that line and shake the ground. You record the seismic waves and look for the squiggle that says "Oil!"

Oil exploration is fine with me. I even have a soft spot for strip mining because it exposes so much rock. Sometimes you find fossils nobody suspected were there. I do like forests and grass and living animals, but I like fossils, too. The trouble in this case was that I already knew there were a lot of fossils in this general area. The seismic line that the surveyors had laid out went right through prime fossil territory, and I was concerned that some good material might be destroyed by the simulated earthquakes. I had some crew members walk along the seismic line wherever it was on Peebles land just to see if they could find anything. Fran Tannenbaum was tired, so to avoid extra walking she took the section of the surveyor's line that would lead her back to the cars.

One of the surveyor's flags was right on top of the small hill we had been walking over and around each day. Fran trudged up to the top and looked down at the base of the flag. Lying exposed on the surface, an inch or two from the point where the flag had been driven into the ground, was a whole dinosaur egg. I heard Fran yell, and I ran up to the top of the hill. The rest of the crew could hear the shouting, so they came, too. Fran was so excited that she was close to tears. Her egg was sitting there in the rock, as plain as could be. Immediately we all got down on our hands and knees, the whole crew, and started looking for eggs. They were just lying all over the ground. All you heard was "Got one," and from somebody else "I got one" and another "I got one, too!"

This was still our first season, and we had found maiasaur nests only in the first nesting ground. None of these contained any intact or partially intact eggs; we didn't even know at the time what the size of the maiasaur egg was. The reason, we suspected, was that the young dinosaurs were trampling their eggs while they jostled each other in the nest waiting for food. These new eggs were remarkable not only because they were intact eggs, but because they were obviously from a different kind of dinosaur. The eggshell had a different surface—not ridged, but pebbly, or bumpy. And, as we would learn later when we found intact maiasaur eggs the size of elongated grapefruits, the ones we were finding on the hill were smaller and narrower, about four inches long. They didn't seem to be laid in nests, either, or in clutches. They were found on their sides in rows. There was no line of demarcation that we could find showing a hollow or any other preparation for the eggs. That first day we found five eggs, some bone, and shell fragments.

We went to Choteau the next day for shopping and showers, and I called Princeton and the office of the seismic testing company to ask them to go around our spot. In another day or two, a Shell Oil executive flew up from Texas to look at this dinosaur site that was in his way. He came out, walked around, looked at the hill and the fossils we'd taken from it, and went home. When the seismic crew arrived a week later with their explosives, they put a little jog in their straight, surveyed line and blew their hole on the side of the hill where it didn't matter to us.

WE APPROACHED THIS NEW SITE, which came to be named Egg Mountain, with great care. When you look for small fossils, eggshell, or fragments of young or even embryonic individuals, you have to treat a chunk of rock differently from the way you do if you're looking for sauropod femurs. Paleontologists who hunt for very early mammalian fossils have the worst time of it. They sift through tons of dirt and rock, looking for bones the length of a fingernail and the diameter of a toothpick. We don't always have it that much better. Some of the fossils we search for aren't more than an inch or two long.

To begin with, we mapped an 8,400-square-foot section on the west face of the hill, where the first eggs had been found. We did this by laying out, with string and stakes, a grid of 21 squares, each 20 feet on a side. The map was created on a scale of one page for each square of the grid. For each square we followed the same procedure. First we picked up the obvious bones on the surface, recording their locations on the map. For every succeeding step we would also record all fossils in their proper location on this multipage map. After the initial collection of bone fragments, we used an ice pick and brush to take out any bones stuck in the dirt or crumbling out of the surface of the rock. Next we gathered up all the loose dirt and rock and sifted it through a screen. The dirt fell through, leaving pebbles and pieces of bone on the screen to be picked over. Finally, when we'd gotten the dirt off and were down to hard rock, we got on our hands and knees again to wash and scrub the whole top of the hill with water and stiff brushes.

We took one whole season to get down to clean rock and sift through everything. Of course, we were also digging up the hadrosaur nesting ground and parts of the big bone bed. In the process of cleaning off the top of the hill we found more eggs, of still a third sort. These eggs were about six inches long. They had one pointed end and one blunt end. They were smoother than the pebbly/bumpy eggs that were the first fossils we had found on Egg Mountain. They were also different from the crinkly-surfaced maiasaur eggs from other sites (there were no maiasaur eggs on Egg Mountain). The only interruption of the egg surface was a series of barely visible, very thin striations, or stripes, that ran lengthwise. These eggs were found partially intact, laid in a spiraling circle. One clutch was on one level, or fossil horizon, and another clutch was on a different level. The eggs had been preserved standing up, vertically, with their bottoms embedded in the hard rock, so that they looked as if they had been stuck in the ground when they were laid. We also found bones, including pieces of skull, and a tooth that looked like it belonged to a small dinosaur called *Troödon*.

At the time we were quite confused about who had laid what. The bones and the skull that we had found appeared to belong to some kind of hypsilophodontid, a small ornithischian dinosaur about the size of a German shepherd. The hypsilophodontids existed, with some variation in form, for the full 140 million years that the dinosaurs lasted. Our guess was that they had laid the spiral clutches of eggs, partly because they were ornithischians. In previous discoveries of dinosaur eggs, ornithischians had produced eggs in clutches. The maiasaurs and *Protoceratops* from Mongolia were both ornithischians; both had produced eggs in clutches. The few other dinosaur eggs that had been found, for instance in southern France, were laid not in nests or circular clutches but in straight lines, as the pebbly/bumpy eggs seem to have been. These eggs were all thought to have been laid by saurischian dinosaurs. In France, in particular, fossils of saurischian dinosaurs were found on the same horizon as the eggs. So our guess was that the hypsilophodontids had laid the clutches. We had no idea who had laid the pebbly/bumpy eggs. (And we still have no idea.)

Despite the confusion, we were certain of two things. We knew we had egg clutches laid by a dinosaur other than *Maiasaura* and that these egg clutches were on two different horizons. Our first thought was that perhaps we would find more egg clutches if we could uncover these two horizons. The question was how to go about it. The next two seasons, while we were pursuing our other work, we tried to figure out what to do with Egg Mountain. We looked at the hill, walked around it and over it, and poked at it occasionally as if it were a ship from outer space that we weren't sure how to open. We thought about the puzzles it offered us. Who had laid what eggs? What kinds of dinosaurs were they? We combed other parts of the hill for fossils, and on the south side, where we had not scraped, swept and scrubbed, we began to find more clutches of what we were now thinking of as hypsilophodontid eggs. (These were the 15-centimeter, pointy-ended, smooth eggs in circular clutches.) We found four more of these clutches. By the end of the 1980 season, we had a total of six clutches of the smooth eggs and six individual pebbly/bumpy eggs that seem to have been arranged in paired lines. We had found each kind of egg on each of three different fossil horizons.

These horizons were essentially two-foot-thick layers of an unbelievably hard mixture of calcium carbonate and silica, a kind of limestone, something similar to the calichi layers found in the maiasaur nesting ground but thicker and harder, and not broken up into nodules. Egg mountain was all rock, and most of it was this calcium carbonate with some shale layers between it. When we found the eggs, they were surrounded by shale with their bottoms stuck in the calcium carbonate. After we had finished our excavation, we figured out what had been happening: Egg Mountain, in the time dinosaurs frequented it, was apparently a peninsula or island in a shallow alkaline lake. When dinosaurs came to lay their eggs there, the level of the ground was almost the same as the level of the water—it was a low island in a low lake. Dinosaurs probably laid eggs and walked around on that surface when it was muddy enough that their walking disturbed the soil. We can see the bumps and wrinkles of the soil surface preserved in the calcium carbonate. However, it could not have been soggy mud, because the eggs would rot in soggy mud. Or they would sink into it and the embryos inside would suffocate. One thing eggshells do is allow oxygen in and carbon dioxide out. Creatures that need oxygen once they're hatched need oxygen while they're in the egg, too.

After the lake dried out, the muddy soil, which was saturated above and below the water line with silica and calcium carbonate, turned into dry hardpan soil, which eventually turned into this limestone, like calichi except that there don't seem to be nodules. In addition, when the lake was dry, streams and creeks ran through it, depositing muck in the spring floods and burying the eggs. This muck turned to the shale that we see around the top of the eggs. Even before we'd begun to take Egg Mountain apart, we thought these different horizons might each represent a nesting season. For one thing, we could see the trampled, muddy soil surface clearly preserved. And all the eggs were at the same depth in the limestone.

It seemed likely that we would find more nests if we could follow these several horizons. We were assuming that these hypsilophodontids, like the maiasaurs, gathered in social groups to lay their eggs. But the limestone was incredibly hard. With pick and shovel, we made very little progress. So, in 1982, we brought in a jackhammer and began to chop the top off Egg Mountain.

This was not standard practice in paleontology. Usually paleontologists dig where they know there are fossils. They don't expend energy to uncover deposits that they think *might* hold fossils. Nonetheless, we had done something similar before. During the 1981 season, we had tried to expose a portion of the lake bed that surrounded Egg Mountain. This was a little bit of shale peeking out of the side of a hill so steep that it was almost like a cliff. I knew that we could follow this shale deposit, which was the preserved muddy bottom of the lake, if we went into the hill. I had it in my head that if we did this we might find some dinosaur footprints. For two weeks a crew of three, Bob Makela, Wayne Cancro and Pat Leiggi, spent all day with a jackhammer and a couple of shovels, driving down through about 15 feet of rock to expose more of the lake bed. It was a good crew. None of them minded the work. In fact, Pat was so entranced by paleontology that he went on to become a preparator. He worked for me at Princeton, he worked at the Philadelphia Academy of Natural Sciences and he is now my chief preparator at Montana State University.

We did find the lake bed that summer, but no footprints. That was a lot of work to do without finding what we wanted. However, we found other things. We found small stromatolites, solid little domes of blue-green algae four to five inches in diameter and two to three inches high. Blue-green alga is one of the earth's oldest lifeforms, having been around for two billion years. If blue-green alga could think and observe what went on around it, the dinosaurs would have seemed like a 140-million-year flash in the pan. We also found clams and snails, and cracks in the mud, which meant that the lake dried up at some point. We found plants called charophytes, a kind of green alga that is known to live in shallow, alkaline lakes. Later we were able to follow this lake deposit all over the anticline. We estimated that the lake covered perhaps 10 square miles and that it surrounded the nests we were finding on Egg Mountain. This wasn't the information we'd been looking for when we started out, but still it was very valuable. Despite or perhaps because of the experience of going after the lake bed, we resolved to go after Egg Mountain the next summer.

In 1982 we began. We cut straight down into the hill. The limestone was weathered on the top of the hill. Wind and rain had already made the initial attack on it, causing it to crumble in spots and develop cracks. Because of the advance work done by the weather, we could go straight down into it with the jackhammer, at a 90-degree angle. As we progressed in later years to limestone that had never been exposed to the weather, the going got tougher. When the going gets tough, you change your tools. We got a bigger jackhammer, with a compressor big enough to have been built with wheels and a trailer hitch. Unweathered Egg Mountain limestone was so hard that if you tried to cut straight down into it, even with the big jackhammer, the blade just bounced off the rock. Once we'd gotten the top horizon exposed and no weathered rock was left, we had to cut into the hill from the side, on a diagonal, to try to expose the other two horizons that we knew were there.

The person who supervised the assault on Egg Mountain was Bob Makela. He was uniquely qualified for the project. In addition to his paleontological knowledge, he could handle a jackhammer instead of having the jackhammer handle him. He also saw in one intractable problem an opportunity to indulge his passion for carpentry, a passion that by now had become obvious in his construction of the camp itself, with its root cellar, kitchen, loading platform, and tire-tube solar shower.

The problem was how to get 100 or so tons of rock off the top of Egg Mountain during the excavation. We didn't want to have to dump the rubble on a part of a hill that we would excavate later, because then we would have to move all that rock twice. Bob solved the problem by building a chute that looked something like a giant amusement park water slide. We dumped the rock on the chute and then pushed it down, sliding it to the bottom of the hill, out of the way. Sometimes the crew members would stand on the chute to push the rock down and in the process slide down behind it. They began to talk about skiing Egg Mountain, which is how the site got its name.

The work went like this: Bob would start with the jackhammer and go along 10 or 20 feet on a line, breaking out big pieces of rock. Then he would stop and stand up on the hill for a while, watching everybody else work. Below him, kneeling on the ground, five or so people would pick up each chunk of rock and look it over for fossils. Usually what you would see would be, not the broad surface of an egg, but rather an edge of broken eggshell that appeared as a thin black line in gray rock. The bones of hypsilophodontids were a bit easier to see, but they were also small and required close examination of the pieces of broken rock. Any fossils got turned over to Bob. The kneeling prospectors would break up each chunk of rock into smaller chunks with a rock hammer, until they were satisfied that if there were an egg or part of an egg in there they would have seen it, or until it seemed futile to break it up any longer. The prospectors would toss the rocks behind them when they were done with them.

Then one or two other workers would shovel the rejected rock debris onto the chutes and off the mountain. There was a tremendous amount of debris involved. In 1982, the first summer of the jackhammer, we moved 25 or 30 tons of rock off the top of the hill. That's maybe 60,000 pounds of hand-chopped, closely inspected bits of limestone and shale.

The process wasn't easy on any of the workers. There are no trees on Egg Mountain. There was grass when we started, but we got rid of that fairly quickly. While we were working there, the hill had no vegetation. Just rock, sun and us. The humidity on a hot day in the summer, when the temperature is 100, may be as low as 10 percent. Supposedly that makes the heat easier to take, but when you're working in it you lose body fluids rapidly. And if you don't have a hat, you're courting heat stroke. Bob would bring coolers of soda and beer to the site, and run the radio in his jeep on a rock-and-roll station while everybody worked. In the middle of a cattle range there would be five or six people on their hands and knees, looking for dinosaurs to the sound of electric guitars.

When the crew did run into a nest, the jackhammer was not retired. Usually a nest with eggs, like the nest of hadrosaur eggs that got the helicopter lift out of Egg Gulch (a.k.a. Egg Dance Coulee), is treated very gently. We expose it with rock hammers, ice picks, whisk brooms and toothbrushes until we can see the extent of it. Then we chisel under the clutch until it sits on a pedestal like a rock toadstool, coat the whole nest with burlap strips soaked in plaster, let it dry, and then break the pedestal, lifting the nest out whole and nicely protected. We did a similar thing at Egg Mountain, but not so gently. Bob used the jackhammer to pedestal the nest.

How we handled the fossil material once we got it chopped out of Egg Mountain was in dramatic contrast to the way we attacked the mountain. We had to chip away at the chunks gently to expose the eggs and the small bones of the skeletons. When we got one of the few partial jaws or the one prized skull complete with jaw, we had to proceed even more carefully. One way of doing this is to use a needle, under a dissecting microscope, to expose the small bones. Another is acid preparation.

Jill Peterson did most of the acid preparation at the dig itself. We had devised a small laboratory in an office trailer with a couple of plastic trays for setting up acid baths for recalcitrant fossils. Jill did a lot of work there on the fossil material from Egg Mountain. Acid preparation is a process designed for dealing with limestone. The underlying principle is that limestone, no matter how hard it is, can be dissolved by acid. This weakness of limestone (and other stones, like marble) is the reason that acid rain and atmospheric pollution are destroying Greek ruins, statuary throughout the world, and the façades of buildings.

When a small fossil is embedded in limestone, the result of the stone's disintegration is more gratifying. First you paint a protective coating on whatever bit of bone you can see. Now we use an acetate-based glue, but in the beginning we used a mixture of benzene and styrofoam. Benzene is not something you want to breathe a lot of, and we stopped using it before the Occupational Safety and Health Administration caught up with us. The next step is to immerse the whole chunk—bone and painted fossil—in an acid bath. This sounds grim, as if the flesh would fall from your hands if you put them in the acid, but limestone dissolves with very weak acid. And the goal is slow dissolution. Something like hydrochloric acid is so strong and works so fast that it ends up destroying the bones, too. We use it only when we have to dissolve a big chunk of rock covering a fossil, and we let only the rock into the acid. Mostly we used a mild acetic acid when we first started doing this kind of preparation. On the most delicate bones we use citric acid, at about the strength of lemon juice.

Once the exposed bones are painted with glue and the fossil-bearing rock is in the acid bath, you wait while the acid works. This can take anywhere from an hour or so to all night, depending on the strength of the acid and the amount of rock you want it to eat away. Then you take the chunk out and rinse it off in water. You inspect it, and you look for newly exposed bone. You paint all the bone with glue again. You put the rock in the acid again. And you keep doing the same thing over and over until you get the bone exposed well enough so that you get a good look at it, whether it's an inch-long embryonic tibia or a jaw with virgin teeth never used to chew because the dinosaur the jaw came from never made it out of the egg.

EACH YEAR WE CHOPPED at the mountain and then picked at the results. In 1983 we took about 50 tons of rock off the mountain. And we did the same thing again, tackling another horizon, in 1984; that year we moved only about 30 tons. The result of our work, in new discoveries, was about one nest per season. By the time we were done, we had found two different kinds of dinosaur eggs, 25 hypsilophodontid skeletons, 12 clutches of hypsilophodontid eggs, teeth of hypsilophodontids, carnivorous dinosaurs and a third kind of dinosaur that we had some trouble identifying, and mammal, lizard and insect remains as well as root casts of plants. The most important of these finds were the smooth eggs, the ones we believed came from some variety of hypsilophodontid. We found the 12 clutches of these eggs on three horizons. There were four clutches on the top horizon, five on the middle and three on the bottom. All the clutches were the same, the bottom halves of eggs arranged in spirals embedded in calichi. On each horizon were scattered skeletons of young hypsilophodontids. We judged that at maturity these dinosaurs reached 8 to 10 feet in length. And the nests were 8 to 10 feet from each other; they were separated by the length of an adult, as the maiasaur nests had been. With all three horizons and 12 nests, the evidence seemed incontrovertible. We had found our second variety of dinosaur—the hypsilophodontids—that had gathered in colonies to lay its eggs.

At first, however, we didn't know exactly what kind of dinosaur was laying these egg clutches. We didn't know it was a hypsilophodontid. At first we thought the producer of the egg clutches was the dinosaur called *Troödon*.

Like many other varieties of dinosaur discovered and described in the early days of paleontology, the original report of this creature

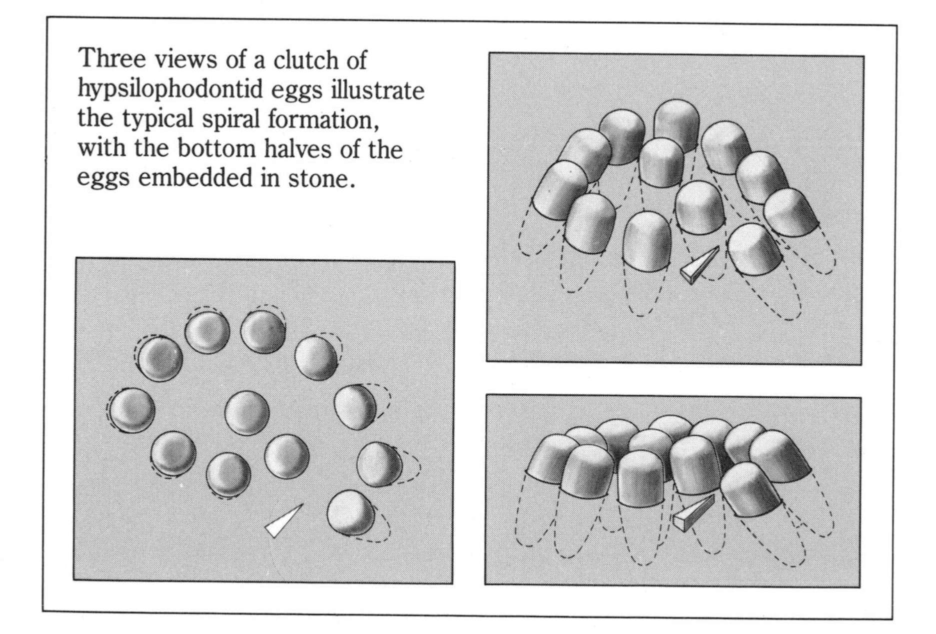

Three views of a clutch of hypsilophodontid eggs illustrate the typical spiral formation, with the bottom halves of the eggs embedded in stone.

was based on one tooth. It so happens that the tooth was one of the first dinosaur fossils found in North America. Ferdinand Hayden found the tooth in 1854 in the Judith River formation. It was in that first batch of dinosaur fossils found in North America, along with the hadrosaur tooth I mentioned before. Now, on the basis of one tooth it wasn't easy to say precisely what kind of dinosaur *Troödon* was, but Joseph Leidy at the Philadelphia Academy of Natural Sciences named it *Troödon formosus* when he described it in 1856. He did not think it was a dinosaur.[1]

Later paleontologists classified *Troödon* first as a saurischian, then as an ornithischian and then back again as a saurischian. Other teeth and skeletal fossils that later turned out to be from the same dinosaur were found along the way and given their own names. The situation was one of great confusion. About all that was clear was that *Troödon* was a carnivorous dinosaur, because its teeth seemed obviously designed for tearing flesh. What happened on Egg Mountain was that, as we were first collecting fossil material, we found *Troödon* teeth and hypsilophodontid skeletons. Sometimes these skeletons did contain partial jaws with some teeth. The teeth weren't the same as the *Troödon* teeth, but hypsilophodontids are very primitive dinosaurs and one of their primitive characteristics is that they retained, even in the Cretaceous, two kinds of teeth. They seem to have kept an all-purpose dental setup, of the sort that could be characteristic of an all-purpose dinosaur before it started to evolve into more specialized forms. Hypsilophodontids have grinders in the back of their jaws and carnivorous-looking teeth in the front. We first found jaws with only the grinding teeth present, and we thought a full jaw would show additional teeth—the carnivorous-looking *Troödon* teeth. This did not mean that we had found some cross between a *Troödon* and a hypsilophodontid. What we thought it meant was that there had been some misclassification—that this *Troödon* creature was really a variety of hypsilophodontid. And this would have been quite a strange occurrence in vertebrate paleontology because it would have been the first case of an

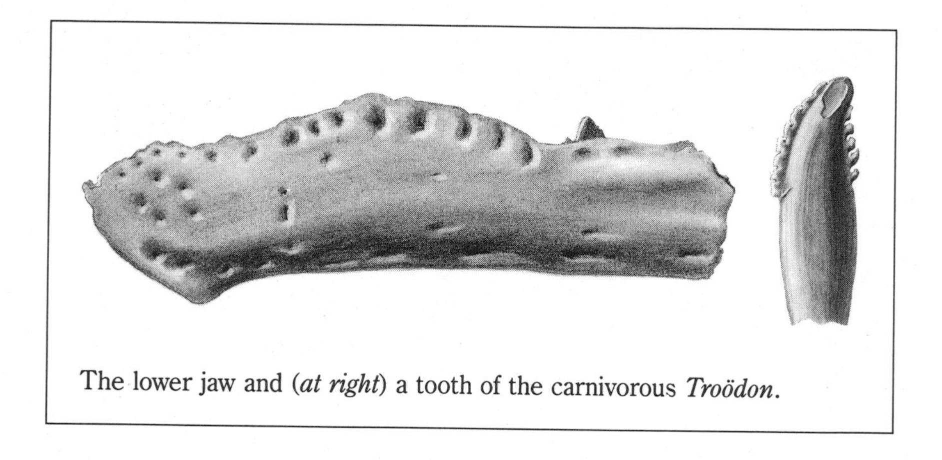

The lower jaw and *(at right)* a tooth of the carnivorous *Troödon.*

apparently carnivorous ornithischian dinosaur. The hypsilophodontids were ornithischians and, along with all other ornithischian dinosaurs, were believed to have been herbivores. All the carnivores were saurischian.

In 1983 I was visiting Phil Currie in Drumheller, Alberta, looking at the site for the new Tyrrell Museum of Palaeontology. We were walking around the site, talking about Phil's work on small carnivorous dinosaurs, when I saw a fossil on the ground and stopped to pick it up. It was a piece of the jaw of a small carnivorous dinosaur. We knelt down to look at where it came from, and we could see that the rest of the jaw, which in total was three or four inches long, was going into a hill. Phil decided to bring his crew back later and dig it up. At the time he was quite excited, because it seemed that the jaw had the tooth of a thing called *Stenonychosaurus,* and *Stenonychosaurus* was thought to be the same animal as *Troödon*. It was just one of those cases where several different finds had been given different names but were all thought to be the same creature.

When Phil's crew went back, they couldn't find the site. When I was there again, two years later, we poked around and I found the spot. This time we marked it carefully, and it turned out, after the crew got the jaw out of the ground, that it clearly had *Troödon* teeth of the sort we'd found on Egg Mountain. Comparing this jaw to all the other fossils of various names but with similar teeth led Currie to conclude that *Troödon* was the original name for at least three different genera and should therefore be the one that stuck.

The jaw also solved a puzzle for us, because it was clearly not a hypsilophodontid jaw. It had different kinds of teeth in it, but they were all carnosaur teeth—sharp, meat-rending. *Troödon* remained what everybody thought it was: a small, carnivorous, saurischian dinosaur. It was not a hypsilophodontid. On the contrary, it probably ate hypsilophodontids, at least the small ones. And it left its teeth, and nothing else, on Egg Mountain because, like all dinosaurs, its teeth fell out periodically and new ones came in. When would teeth be more likely to come out than in the process of catching, or eating, prey?

So, what dinosaur had laid the eggs? We were now able to exclude the *Troödon* teeth from consideration and to look only at the skeletal material and the hypsilophodontid jaws. Obviously we had some kind of hypsilophodontid. It wasn't until 1984 that we finally got a complete skull with jaws and teeth. This one took Jill almost a year to prepare, using a combination of acid baths and scraping away at the delicate bones with a needle under the dissecting microscope. It finally revealed to us the two varieties of teeth common to hypsilophodontids: short, triangular teeth and sharp, cone-shaped, slightly recurved teeth that, though not as carnivorous-looking as *Troödon* teeth, certainly seem as if they were designed to tear flesh. Such a dental setup is reminiscent of the very first ornithischian dinosaurs, the fabrosaurids, which appeared in the late Triassic—more than a hundred million years before these hypsilophodontids laid their eggs at the Willow Creek anticline. The hypsilophodontids we found were also slender and would probably have been very fast runners. Again, in this they were very much like the fabrosaurs. What this meant was not that the hypsilophodontids were throwbacks, but simply that they were evolutionarily conservative. And, of course, they shared their upland areas with other conservative dinosaurs, the maiasaurs. Dave Weishampel and I worked on the description of this dinosaur, and we called it *Orodromeus makelai* in honor of Bob Makela and his work on Egg Mountain.[2]

I SHOULD POINT OUT that though we solved the identity problem of the new dinosaur, we didn't solve the problem of what it ate. It is assumed that all ornithischians were herbivores. And certainly the hypsilophodontids were ornithischians. But they had teeth that could have been used for almost any purpose. I don't see why they couldn't have been omnivores or, for that matter, insectivores.

Having identified our nesting dinosaur, we were able to reconstruct what you might call a day in the life of Egg Mountain. First we must imagine it not as a hill at all but as a very low-lying island in a shallow, milky green alkaline lake, something very much like the lakes called playas that now exist in arid environments in the American Southwest. The scene is not one that is full of life, with shore birds wading and little fish darting about. It is the lake whose bed we had uncovered in our first effort with the jackhammer. It was big, but so heavily alkaline that not very much lived in it. There were no fish, but there were clams, snails and colonies of algae that built solid structures in the water looking like rock toadstools three or four feet high. We know that microplankton lived in it, and perhaps some crustaceans such as brine shrimp.

We also found, in the lake sediments, the fossil remains of a very large pterosaur. It had died and been entombed in the mud of the lake bottom. But we have no idea what it was doing there. Perhaps it was just flying over and made an emergency landing. Perhaps birds fed on some microscopic lake creatures and the pterosaurs fed on the birds. Or perhaps the pterosaur was a carrion-eater and fed on any carcass it could find. Part of the reason for our confusion is that nobody knows how the pterosaurs made a living. They were not, I should point out, dinosaurs. They were flying reptiles of a variety of sizes and shapes. They all had long bills, and most of them had pretty good teeth. And they obviously ate something, but precisely what remains unknown.

On the northeast shore of the lake we find Egg Mountain. But it is not a mountain; that label refers only to the momentary topography of the Peebles ranch, a result of erosion and the continually shifting earth. Instead it is an island with an area of a little more than an acre. The borders are covered with thick vegetation, which I'm imagining there because of all the root casts we had found in excavating Egg Mountain, so from the shore of the lake whoever is looking sees nothing but waving sedges, or small trees or bushes. The local carnivores cannot readily see the life on the island itself.

It is a flat, low-lying island, not rising significantly above the level of the lake, and if we were to swoop down on it from above, as a pterosaur might during a nesting season, we would see that it was very active. If we came at the right time, we would see hypsilophodontid females laying their eggs. These dinosaurs are slender, quick, and about seven feet long from the nose to the tip of the tail. They don't have the pronounced bill of the maiasaurs, but they do have a long, slender snout. I am certain they nested colonially (or at least gathered to lay their eggs) because the egg clutches on each of the three layers we uncovered were separated by roughly 7 feet, the length of an adult. The maiasaur nests were separated by about 20 feet, also the length of an adult.

But the hypsilophodontids did not have a social life as elaborate as the maiasaurs, nor can we say quite as much about their parental behavior. It seems, for instance, that although they laid their eggs in clutches, they did not make nests; that is to say, they did not construct mounds and hollow them out. They did not build special places to lay their eggs. They did, however, lay them in a spiral clutch, if not precisely a nest. And I'm willing to speculate that they tended these

Egg Mountain: a varanid lizard (*foreground*) is an impassive observer as *Troödon*

chases the young of *Orodromeus makelai*.

eggs, because the clutches seem so carefully arranged. The spacing of the nests also suggests that the adults may have spent time there.

On the other hand, they certainly did not care for their young in the same way the maiasaurs did. Characteristically, the hypsilophodontid clutches show the bottoms of eggs securely in place, still vertical, rather than the pile of crushed eggshell so common in a maiasaur nest. I think this has to mean that the young did not stay in the nest. If they did, the eggshells would have been trampled.

They may, however, have stayed together in a group, in a kind of crèche, the way young penguins do, being cared for by adults in the sense of being shepherded here and there, perhaps even fed. This part of the picture is quite murky and difficult to visualize because of one of the puzzling facts about the fossils found on Egg Mountain. There were skeletons of very young dinosaurs and of older but still juvenile animals. There were no skeletal remains of adults. I suppose that the adults could simply have laid the eggs and left. Perhaps the young stayed because they found ample food (crustaceans or vegetation) in the lake or around the shores of the island.[3]

If, as is possible, they were without adults, they were certainly not alone. The abundance of *Troödon* teeth is the most obvious indication of other life. If we were to fly over the island often enough, taking the pterodactyl-eye view, I'm certain that once or twice we would have seen *Troödon*, a predatory dinosaur probably about the size of a hypsilophodontid adult, snatching either eggs or young hypsilophodontids. We also found some teeth of *Albertosaurus*, a large carnivore that looked very much like *T. rex*. *Albertosaurus* ate the same thing as the proverbial 500-pound gorilla—anything he wanted. Presumably, he occasionally wanted *Orodromeus makelai*, and he occasionally realized that there were a lot of them out on that secluded little island. I don't know how he would have waded out without alerting all the island animals, but perhaps he was stealthy as well as large and fierce-looking.

Had we landed and taken a closer look, we would have seen other things. In the tall sedges there were probably some small mammals, little shrew-like creatures that may have fed on insects. I know there were plenty of insects, because we found fossilized pupal cases in and among all the egg clutches, and my guess is that they were carrion beetles. The insects were most likely scavenging the fluids left in the eggs after hatching and also eating the remains of eggs that did not hatch. Something like a carrion beetle would be likely to enjoy this diet. A newly abandoned nest would presumably have been crawling with beetles.

We would also have seen lizards, probably a foot long, sunning themselves or darting here and there, perhaps stealing eggs. The fossils we have found are of varanid lizards and today among the varanids are the monitor lizards, known for their taste for eggs. With these lizards around it would seem that adult hypsilophodontids would have had to protect their nests, given the high degree of hatching. Most of the eggs we found did hatch, in contrast to the smooth eggs, which were not laid in clutches and were found unhatched. Of course, it may be that these lizards ate something else—insects, perhaps, or small mammals.

I'm confident that this scene repeated itself year after year because we have the several layers, on each of which is a similar remnant of a nesting colony. What seems to have happened is this: When the lake was high (and these kinds of lakes tended to dry up each year) the hypsilophodontids would have come and laid their eggs, and the island would have been teeming with life. The eggs would have hatched, and later the lake would have dried up. The dry bed would, however, still have been laced with small streams that carried sediment, or mud, to bury the old nests. Perhaps the lake itself, when it rose, sometimes went above the surface of the island and deposited sediment on old nests. The dinosaurs would have reappeared to nest again each year, if there was a yearly cycle, or perhaps only when the island was at the right level, neither flooded by a high lake nor left standing as a bump in a dry lake bed. They would have splashed out through the shallows to the small island and laid their eggs in the characteristic spiral clutches.

The reason Egg Mountain is so hard, and we find the eggs embedded in calichi (with alternate layers of the shale that came from mucky sediment deposits), is that the lake was rich in silica and calcium carbonate, probably coming from sediment that had eroded off the rocky mountains, perhaps from limestone that had formed in earlier geologic eras and then been thrust up by the mountains. When the lake was high, silica and the calcium carbonate seeped into the island under the soil surface, which is to say that the body of the island under water acted like a sponge. When the lake periodically dried up, hardpan layers would form not only on the surface but within the subsurface layers of the island.

THERE'S ONE OTHER THING we know about this lake and the hypsilophodontids. There was another nesting colony, on another island. We found it on September 14, 1983, four days before the end of our field season, and this one we called Egg Island. We should have been done with our field season, except that we were waiting for a television crew to come out to do a story. The weather was already cold and windy, and only three of us were left in camp. We took a walk to look at the anticline, a common pastime when there was no actual digging to do—and even when there was. I often walk around and look at things to get them straight in my head, to understand where things are and how they relate to each other and where I might find something else. We looked at some stream sand deposits, the lake horizon and a variety of the anticline's other features. Then, on our way back to camp at the end of the day, just walking along but with our eyes on the ground, as always, we found two more hypsilophodontid egg clutches.

We left the anticline a few days later, and I came back after two weeks to collect the eggs. I got out one clutch with 9 eggs and another clutch with 19 eggs. The big surprise was that the 19 eggs in the second clutch all contained embryonic dinosaurs. I could see the tiny bones near the weathered surface of some of the eggs. Others I cracked open to see if they had embryonic bones. They did. And, to get an even better idea of what was inside, I had them x-rayed, not only x-rayed but CAT-scanned to get a three-dimensional picture. Most of the embryonic skeletons had been disarticulated and had fallen to the bottoms of the eggs, but one was fully articulated.

This was the first embryonic dinosaur skeleton found. We had found a few scattered embryonic fossils of the maiasaurs associated with nests. But in terms of relatively complete skeletons the 19 hypsilophodontid embryos were the very first. There were no other dinosaur embryo skeletons known at that time. A partial embryo, a piece of eggshell with some embryonic bones on it, was considered important enough to warrant a paper in a Russian journal in 1972. And there was a *Protoceratops* egg, displayed for years at the American Museum of Natural History, that was thought to contain embryonic bones. After I'd found the Egg Island embryos, however, I inspected this one carefully, and it was clear that the supposed embryonic bones in this fossilized egg were just calcite crystals in odd shapes.

The embryos were valuable for several reasons. For one thing, no matter how many fossil eggs and eggshells are found associated with fossils of older dinosaurs, it's nice to get final, ultimate proof of dinosaurs in eggs and, in the case of the hypsilophodontids, of which dinosaur was in these eggs. But the embryos have more than that to tell us: we can look at their bone and determine how fast it grows.

Dave Weishampel and I studied those embryonic bones and found that both the maiasaurs and the hypsilophodontids showed the kind of bone that grows very rapidly, the way the bones of modern birds and mammals grow. Furthermore, we were able to compare the state

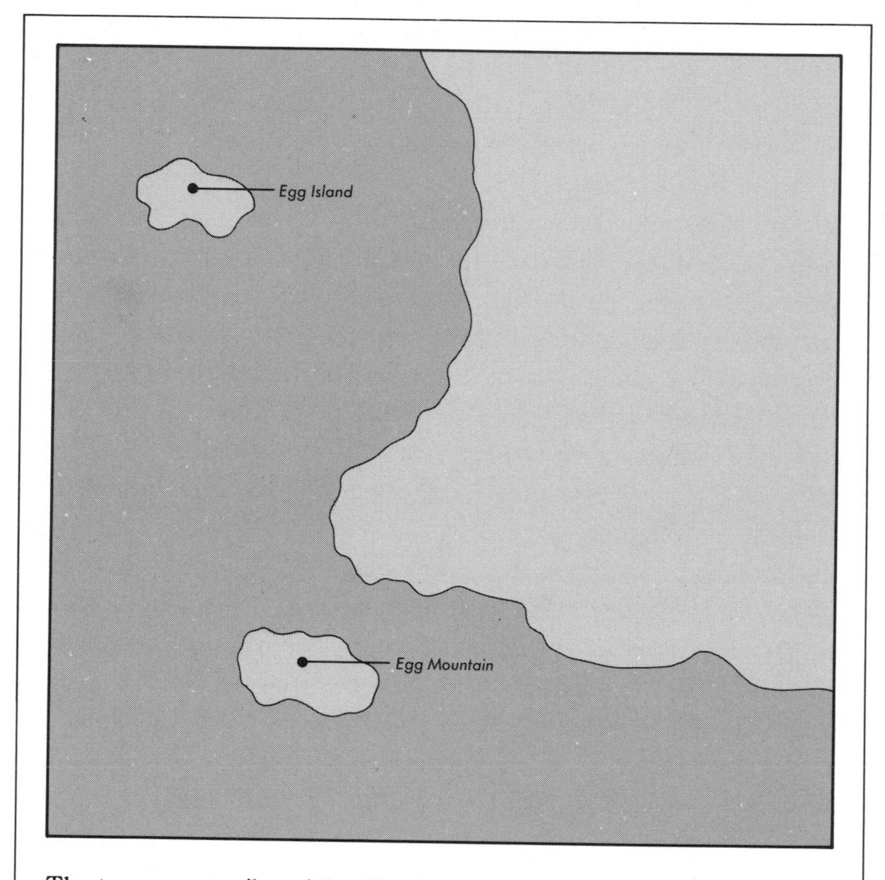

The two square miles of the dig, drawn as they would have looked when Egg Mountain and Egg Island existed. Both features were located in a shallow alkaline lake, represented by darker shadows.

of the embryos of the two different dinosaurs at the anticline.[4] What we found was that the hypsilophodontid embryonic bone was more fully developed than that of the maiasaurs.

Limb bones are formed of ossified cartilage, and one sign of development occurs at the ends of the bones, for instance where the femur attaches to the hip. The end of a bone such as the femur, when it is in fact bone and not cartilage, is called a condyle. The condyle is not there when the bone first grows; at first, the whole femur is cartilage. But as the creature gets ready to be moving about, the condyle ossifies so that the bones connect to each other with hard, bony surfaces. The embryonic hypsilophodontids had well-formed, bony condyles; the embryonic maiasaurs did not. And yet both embryos were of a size that indicated they were just about to hatch. What these facts suggest to me is that each kind of animal was prepared for the life it would face outside the egg. The hypsilophodontids did not stay in the nest after hatching. Even if they stayed together and were watched over by adults it was to their advantage to be quick and relatively mature once they came out of the egg, ready to keep up with the adults and to run from predators. You can move a lot better with well-developed condyles. The maiasaurs, on the other hand, stayed in the nest. Their primary goal in the first month or two of life was growth. They could live in a kind of floppy state with their limbs still largely cartilaginous. If they fell over one another in the nest, so what?

Finally, the embryos completed our set, so to speak. We now had hypsilophodontids and maiasaurs of all ages, from embryo to adult. With all these specimens we could trace the process of growth in the fossil bones, and studying and understanding bone growth is the surest, most secure route to understanding the physiology of dinosaurs. Growth series of fossil bones offer paleontologists the best chance of finding out whether dinosaurs were coldblooded or, as most paleontologists now suspect, warmblooded creatures.

CHAPTER 7

HAUTE BONES

Paleontology is always divided into two seasons. The first is the field season, a time of exploration and discovery, in which you carry your tents with you and search for fossils. Then, at a different pace, is the laboratory season, a time for reflection and study, a season characterized by the tedium of preparation and reconstruction, by the working out of puzzles, the construction of theories and interpretations. The two are inseparable. Without the discoveries, there's nothing to think about. But without thinking about them, the discoveries don't really exist. They are defined by their interpretation. The second season has different demands, subtler pleasures. You might say that the difference between the two is like the difference between Choteau and Paris. That's Choteau, Montana, and Paris, France.

In January 1985, Jill Peterson and I packed our bags with fossils and flew to Paris. We took with us remnants of embryos and juvenile dinosaurs from the Willow Creek anticline to study under the guidance of Armand de Ricqlès, a professor of paleohistology at the University of Paris. Ricqlès knows more than anybody else about fossil bone tissue. In my opinion, his has been the clearest and strongest voice in the debate about whether or not dinosaurs were warmblooded. And he has a scientific mind that—unlike mine, which is the hunt, poke and dig around version usually issued to field scientists—is a model of depth, breadth and clarity.

The trip was the beginning of the second stage in the life of the Willow Creek anticline dig. The first stage was getting the fossils out of the ground; the second was to look at the fossils, study them, make hypotheses based on what we saw and try to prove or disprove them. Paleontology is not an experimental science; it's an historical science. This means that paleontologists are seldom able to test their hypotheses by laboratory experiments, but they can still test them. Two sorts of hypotheses came out of the Willow Creek anticline. The first sort required further field exploration. For example, we had some rather definite ideas about where the rest of the baby dinosaurs were. To test these ideas, we had to go digging in new places, beyond the anticline. We did that in the summers of 1985 and 1986, and I'll describe the results in the final chapter of the book.

But before we did that, we had another variety of hypothesis that we wanted to test. This one had to do with the bones we'd already found. We thought we had, in the fossils of the Willow Creek anticline, enough material to demonstrate once and for all, by microscopic studies of the bone tissue, whether or not dinosaurs were warmblooded.

THE USUAL REACTION TO OUR TRIP, on the part of people who heard about it, was: What is a paleontologist doing in Paris? Well, to give the city, and the French, their due, Paris is not that odd a locale for paleontology. There are fossil shells in the gargoyles of Notre Dame, some of the most evil-looking things I have ever seen. And there is a wonderful collection for some paleoanthropologist in the neatly stacked human bones in the catacombs—femurs in one pile, tibias in another, skulls in a third.

More seriously, Paris is the city where paleontology began and where the first theory of evolution was proposed. Jean Baptiste Lamarck, in 1800, proposed a fully conceived theory of evolution, more than a half-century before Darwin published his *Origin of Species*. His theory was flawed, and in the English-speaking world he is remembered largely for his mistakes. But he made numerous contributions to science, including his studies of fossils of the "inferior animals," as they were then called. It was Lamarck who first described them as invertebrates.

Paris was also the city of Baron Georges Cuvier, another scientific giant. An opponent of evolution, Cuvier lived and worked in Paris at the same time as Lamarck and completely overshadowed him. He analyzed the vertebrate fossils of the Paris basin and was the father of vertebrate paleontology. He saw and reported on the first fossils of ancient reptiles such as *Mosasaurus*, a seagoing lizard (not a dinosaur) whose skull still sits in the Museum of Natural History in the Jardin des Plantes. This is a fossil that was instrumental in changing the whole view of the planet Earth. It was one of the pieces of evidence that made scientists realize that Earth had a long history and that there were animals like the mosasaur that had been on the planet once but had gone extinct. Before these kinds of creatures were found, it was accepted that the species remained immutable, that they did not go extinct, that new ones did not emerge.

Furthermore, in working with Armand de Ricqlès and talking to Philippe Taquet and other scientists at the University of Paris, Cuvier's and Lamarck's university, I got the sense of being in touch with this history, just as in Montana I felt that I was working in the tradition of the early rough-and-tumble fossil hunts in the American West, and just as in Philadelphia, when I put *Hadrosaurus foulkii* back together, it seemed to me that Leidy and I had somehow connected. In Paris, the sense of history is different. The university I teach at is 90 years old; the University of Paris dates back to the twelfth century.

I also got a little taste of Anglo-French rivalry, which is present in science as it is in other fields. When the first dinosaur bones were found, Cuvier was the obvious person to ask for an opinion on just what sort of creature these bones represented. Among the first dinosaur fossils found were those of a dinosaur called *Iguanodon*. The story is that in 1822 the wife of a country doctor and amateur paleontologist named Gideon Algernon Mantell found teeth of this creature. Mantell immediately published a brief description of the teeth (in 1822), although he didn't say what animal he thought they represented.

Cuvier was consulted by Mantell, and when I was in Paris Philippe Taquet had just written a paper on the correspondence between William Buckland, another contender for first discoverer of dinosaur fossils, and Cuvier and between Mantell and Cuvier.[1] One of the reasons Taquet published the paper is that Cuvier is routinely defamed for his failure to recognize that the fossil teeth Mantell sent him were of a new kind of reptile. The standard story, apparently promulgated later by Mantell himself, is that Cuvier thought they were bones of mammals. The correspondence Taquet discusses makes clear that Cuvier has gotten the short end of the historical stick. He wrote to Mantell, with regard to some of the teeth, that they appeared to be the teeth of reptiles but not of carnivores (as all known reptiles were then). And he suggested that what the teeth represented was a new kind of animal, "*un reptile herbivore,*" which, whether you read French or not, obviously means an herbivorous reptile. He did see resemblances to mammal teeth (to rhinoceros teeth, in fact, a resemblance that clearly exists), but he leaned more strongly toward the reptiles.

Taquet's paper does not restrict itself to the identification of dinosaurs. In it is a long, long quotation from a letter that mentioned one of Cuvier's visits to England. The writer of the letter described how inferior English science was (in 1817) to French science. The writer said that in England "the government favors only the art of making money, which is carried, in this country, to its perfection." Long after the relevant historical and scientific importance of the letter has been made clear, Taquet continues the quotation, no doubt to the delight of French readers. The letter writer was of the opinion that (in loose translation) "in England, money has to do with everything, and everything has to do with money." Today I suppose that might be written about the United States.

IN SOME WAYS, this is also true of American science. Largely because of money, the United States has become the world scientific leader. Scientists who want to keep up with research in their field must speak English because the major scientific journals are published in English. In paleontology, North America is also a center, although in this field more because of natural resources than money spent. The American West presents the richest and most varied collection of dinosaur fossils anywhere in the world. In the beginning of the study of dinosaurs, the field research and the interpretation of the bones took place in Europe. However, not long after the first dinosaurs were discovered, the center of paleontological exploration shifted to North America.

And now American paleontologists, and certainly the popular press in America, tend to look to American science for the discoveries, interpretations and stories. I'm sure there is no paleontologist in North America who is not familiar with Armand de Ricqlès. But that's not at all true of the public. When the debate over whether dinosaurs were warmblooded or coldblooded hit full cry in the 1970s, some American paleontologists were catapulted into what, for a dinosaur specialist, passes for fame. In particular Robert Bakker, the most vocal and extreme proponent of warmbloodedness in dinosaurs, came to the fore. To a lesser extent, Bakker's former teacher, John Ostrom of Yale, received popular attention.

Both of them deserve it. They, along with Armand de Ricqlès, were among the first paleontologists to suggest strongly in the late 1960s and early 1970s that dinosaurs were warmblooded. (Ostrom and Ricqlès have both retreated somewhat from their first positions, but not Bakker.)[2] Ostrom has been responsible for a great deal of important research, such as the discovery and interpretation of the fast-moving predatory dinosaur *Deinonychus* that contributed to the reevaluation of dinosaurs. He also worked on the reevaluation of the hadrosaurs and demonstrated, to my satisfaction anyway, that *Archaeopteryx,* the first bird, represents a transitional animal that shows the process of evolution of certain saurischian dinosaurs into birds. The birds, to me, are living dinosaurs.

Bakker, who has taken the role of crusader and gadfly, has made daring conclusions and interpretations about the nature of dinosaurs that have challenged prevailing paleontological thinking. Not everybody in paleontology agrees with Bakker, but I think he has been a spur to more and better research. Out-of-the-way fields such as vertebrate paleontology can grow sleepy and accumulate dust like museum cellars, but not if Robert Bakker is around.

There were many arguments put forth to support the idea that dinosaurs were warmblooded. Arguments were based on the posture of dinosaurs (very much like that of mammals), on the ratio of predatory dinosaurs to prey dinosaurs (again like that of mammals, if the fossil record is reliable for those statistics—something I doubt), on the ecological niche dinosaurs held (that of the dominant terrestrial animals, like the mammals now), and other factors. But the soundest to me always seemed the one based on the hardest evidence (and I mean that literally): the bones themselves.

I first became aware of the bone tissue evidence from reading

Bakker. He recounted studies by two American histologists, D. H. Enlow and S. Brown, showing that dinosaur bone looks more like the bone of mammals and birds than it does like that of reptiles.[3] Enlow and Brown did not make conclusions based on their findings, but Bakker did. He said this similarity was one of the things that showed that dinosaurs were warmblooded. I found out later that Armand de Ricqlès had done his own studies of fossil bones. Ricqlès, like Ostrom, has become more cautious since his first publications and now argues that, though the evidence seems to suggest there may have been warmblooded dinosaurs, there is not enough proof.

Ricqlès is most rigorous about both the technique and the logic of science. Jill and I went to Paris with visions of a quick, clean, final study on bone tissue and warmbloodedness, using the fossils from the anticline. We had a hypothesis, although it was somewhat vague, and we thought that with Ricqlès' aid we would test it and no doubt confirm it. What we got, which was a bit of a surprise but I think of more value in the end, was a three-month lesson in bone histology and scientific discipline.

The trip really began when I sent Ricqlès a sample fossil of the babies we found in 1978 and asked him if he would study them. I had been convinced by seeing the babies and by the geological and environmental evidence that the babies had to grow rapidly. I knew that by looking at bone tissue of the babies, Ricqlès could say, with more clout and certainty than I could muster, whether they were in fact growing fast. Ricqlès' initial papers contended, to oversimplify for the moment, that samples of dinosaur bone showed by and large the sort of bone that comes from the most rapid growth and that rapid growth was often a result of warmbloodedness.

Ricqlès and I carried on a correspondence until, in 1984, he invited me to come and work in his laboratory. I lacked experience in histology but wanted to find out if a histological study would prove that the dinosaurs at the anticline had been growing rapidly and that they were therefore warmblooded. I had been offering to provide the material if Ricqlès would do the work. Ricqlès was interested in a histological study of the bones, but he suggested that I come and work in his laboratory. He would oversee the study and provide the expertise, and Jill and I would do the work under him. This struck me as a great idea. I even tried to learn French, a project that I gave up once and for all when, in an attempt to call Ricqlès to discuss arrangements for the trip, I was unable to get past the Paris operator. She hung up on me. I trudged over to the romance languages department at MSU and got a friend there to make the call. Once he had Ricqlès on the phone, I got on. Ricqlès speaks English, probably better than I do.

To describe what we did in his laboratory, and what we found, I should explain first what histology—in particular, bone histology—concerns itself with. In the body, bone is built somewhat the way sedimentary rock is constructed, by deposition. To do the histology of fossil bones is to do a kind of microscopic sedimentology. Rocks are deposited and eroded, and we look at the record of deposition and erosion to read the history of the rock formations. Bones are built in the same way. Calcium is deposited, eroded, redeposited. And histologists studying bones work much the way geologists do, reading the history of bone growth from clues left in the finished bone. They can also read the history of disease and injury in fossil bones.

The bones that are important in ferreting out the secrets of how an animal has grown are the internal, or skeletal bones. There is another kind of bone, called dermal bone, from which the scales of fish and the skull plates of cats and human beings are formed, but it doesn't have the kind of structure that leaves clues about growth. The skeletal bones in higher vertebrates do have these clues. They are formed by the transformation of cartilage. The whole skeleton, in the embryo of a lizard or a person, starts out as cartilage. As the embryo develops, this cartilage turns into bone.

One of the characteristics of both cartilage and bone is that it consists of living cells embedded in nonliving material. The cells secrete a matrix that they rest in. In cartilage, this matrix is flexible and gelatinous; in bone, the matrix is rigid, composed primarily of calcium in one form or other. As the cartilage turns into bone, on the outside of, say, the femur, calcium is deposited in sequence, the way sedimentary rock is made. (Bone is deposited from the inside out, while rock is deposited from the bottom up.) The deposited calcium, called perichondrial bone, is the bone that we're concerned with.

The point of most importance here is that the structure of the bone is directly related to how fast the bone grows. The fastest-growing bone is densely riddled with canals for blood vessels (Haversian canals), and these canals go out in all directions. It does not show lines where growth has stopped (growth rings) and is not deposited in clear layers. This kind of bone is called plexiform bone. The slowest-growing bone has fewer canals and shows growth rings, like those in a tree, that mark times in the animal's life when growth stopped or almost stopped; these growth rings are very common in coldblooded animals because they grow more during warm seasons. That kind of bone is called lamellar-zonal bone.[4]

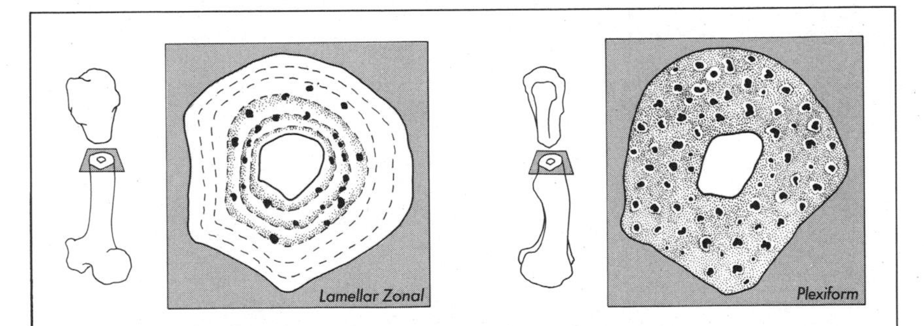

A simplified diagram of lamellar zonal bone (*left*) with growth rings and relatively few Haversian canals (dark spots). Plexiform bone (*right*) has no growth rings and many canals.

Naturally, in order to see the interior structure of a bone, you need to cut a cross section. And since the details we were looking for could be seen only under a microscope, our cross section needed to be thin enough to go on a slide under a dissecting microscope so that light would shine through it and illuminate the bone's interior.

The first step was to take a bone, let's say a piece of a hypsilophodontid femur, and coat it with a resin so that it became like an insect in amber. The purpose of the resin was to hold the bone together so that it wouldn't shatter when it was cut. The next step was to slice off a cross section on a circular saw. We then glued the cross section onto a piece of glass to make the beginning of a microscope slide.

Once we had the section on glass, we had to grind it down. We held each glass-backed section on a flat, circular grinding wheel, something like a potter's wheel. As it turned, water from a hose dribbled onto it, and we applied an abrasive substance, ranging from powdery to sandy in consistency, to aid in grinding. The problem with this procedure was that we had to hold the section down with our fingers. Gloves were impractical because the glass slides were often thin and we needed the sensitivity of skin to feel that we were holding them in place. This very sensitivity, however, posed other problems. Our fingers often slipped. The grinding process was slow, so there was no sharp abrading of the skin when our fingers touched the grindstone. But gradually, after many mistakes, the stone wore down the skin, so we ended up with bleeding fingertips. The blood didn't run; it just sort of oozed. In the first month, we did 100 slides. All told, we did 360. I recall times when Jill, who did more of the grinding than I did, would have to stop because she didn't have enough intact fingertips to hold down a slide. Seven or eight of her fingertips would be bleeding. When it came time to check that day's work under the microscope, she would have little bits of tissue to absorb the blood (the kind men use when they make a mistake shaving) on the tips of most of her fingers.

We cataloged each slide that we made. Then we examined it under a microscope and recorded a description of the cross section in a notebook. First, of course, we looked to see if we'd gotten the slide thin enough so that light would shine through the cross section and illuminate the internal structure of the bone. Then we had to learn to see what we were looking for. Just as untrained eyes don't notice the difference between bits of fossil bone and bits of rock, or don't see an anticline or a deformation of sedimentary rock beds, untrained eyes don't recognize Haversian canals, lamellar zonal bone and plexiform bone. With Ricqlès' help, we trained our eyes.

It was a bit like learning French, but much easier for me. From a textbook, you can learn words or phrases or sentences. But speaking French with a Parisian, having a real conversation with all the imperfections and variations of real sentence structure, idiom, enunciation and accent, is very different from reading a textbook. In the case of bone structure, looking at a real maiasaur tibia is very different from looking at diagrams or reading textbook descriptions. You need a guide, an interpreter, someone to show you the way—to show you what "conversational" histology is like. Ricqlès was our guide.

As a guide, however, he led us where *he* wanted to go. Surprise number one was that my notion for a study of the bones, my plan for confirming the hypothesis that these creatures were fast-growing and warmblooded, wasn't going to work. First of all, I had planned to use the bones of embryos and young because there had been some suggestion that the size of an adult dinosaur could affect its bone. It might have required lots of vascular canals if it was quite big, whether or not it had internal temperature regulation. So I thought we could solve that problem by studying the youngest dinosaurs. Ricqlès informed me that such a study would be misleading because in all animals, cold- or warmblooded, extreme youth is the period of fastest growth. It was necessary to study a full growth series of a given animal, from embryo (if possible) or hatchling to adult.

Second, Ricqlès pointed out that even if we had a record showing that a given animal grew rapidly, this would not necessarily mean it was warmblooded. It would certainly suggest warmbloodedness, but what it would really mean was simply that the animal's metabolism was high enough, throughout its life, to keep a fast pace of growth. How, other than being warmblooded, could an animal keep the internal fires stoked hot enough to grow rapidly? Well, to answer that question, it's necessary to inspect the whole notion of warmbloodedness and coldbloodedness, which is not at all as simple as it sounds. In fact, the terms "warmblooded" and "coldblooded" are, for scientific purposes, so vague as to be almost meaningless.

Temperature regulation is managed, in living animals, not in two ways but in a number of ways, and there are a number of scientific terms to describe them. The terms sound like jargon, and I know there is a tremendous impatience on the part of nonscientists with the long, Latinate words that seem to make a subject more complex than it need be. But some subjects are complex. And if you want to think clearly about a complex subject, and not just fool yourself with words that are easier on the ear, you need language that reflects that complexity.

When we call animals warmblooded, we usually mean that they keep a steady, high internal temperature regardless of the outside temperature. And when we say an animal is coldblooded, we usually mean that it has a relatively low internal temperature that changes in response to the outside temperature. There are several distinctions included in each of these definitions:

1) How body temperature is regulated.

2) Whether the internal temperature is constant or varies.

3) Whether the temperature and metabolic rate are high or low when the animal is at rest.

To describe the first distinction, scientists use the terms "endothermy" and "ectothermy." *Endo-* means "inside" and *ecto-* means "outside"; *thermy* comes from the Greek *thermos*, which means "heat" or "temperature." An endotherm has an internal regulatory system that maintains its temperature at a constant level. An ectotherm has a body temperature that responds directly to outside temperatures. All animals, however, including ectotherms, regulate their own temperature. None reflects the ambient temperature with pure passivity. A typical coldblooded chameleon will move into the sun or the shade to change its body temperature. Even though it's dependent on ambient air temperature, the chameleon has, in its habitat, a range of temperatures to choose from.

The second distinction has to do with the result of regulation—a constant high temperature or a normally low but widely varying temperature. (No vertebrate has a constant low body temperature, because this would not allow much activity.) "High" and "low" are relative terms, and generally "high" refers to the range of resting body temperatures that are characteristic of birds and mammals, and "low" to the much lower resting temperatures that are characteristic of reptiles and amphibians. Maintenance of a constant high body temperature is called homeothermy. And the wide variation of body temperature in response to environmental temperatures is called poikilothermy.

The third distinction has to do with metabolism. A simple definition of metabolism would be that it consists of all the cellular activities and chemical reactions that transform food into energy and use that energy to run the body's activities. Endotherms maintain their temperature at a high level by having a high metabolism; in other words, all that cellular and chemical activity occurs at a fast pace. The ectotherms we know have a much lower resting metabolism. Animals that keep their metabolic rate high all the time are tachymetabolic. The ones that have a slow, or low, resting metabolism are bradymetabolic.

It is true that, in general, an endotherm is also a homeotherm and tachymetabolic. And, in general, an ectotherm is a poikilotherm and bradymetabolic. But not always. You can mix these three categories together and find some animal that fits almost any combination. Think about bats. Most bats lower their metabolism at night, or when food is not available. So, although they're endotherms, bats are not exactly homeotherms and they don't stick to being tachymetabolic. Some hummingbirds are similar: they drop their metabolism to a reptilian level at night when they can't feed; furthermore, when they let their temperature and metabolism drop, they come close to the ambient temperature, thus flirting with real ectothermy. What are they, then? Some people call them heterotherms, which means that in temperature regulation they're switch-hitters.

And what about the Australian monitor lizard? Obviously coldblooded—a bradymetabolic, ectothermic poikilotherm, right? Well, not exactly. The Australian monitor does have a low resting metabolism, but it has some interesting options added to the basic model. It can shift blood from its limbs to its body core when the outside temperature is dropping. The result of this action is that, although its limbs will reach the outside, cooler temperature in about 15 minutes, its body core may take up to 7 hours to cool down. That's approaching a kind of endothermy. Some sharks and tuna can keep a body temperature higher than that of the surrounding water, as some snakes can keep a temperature higher than that of the air. The large leatherback turtle, to take an even more surprising example, can keep the temperature deep in its body up to 18 degrees centigrade above the water temperature; it seems to do this by metabolic changes and body fat insulation. Even some insects and some plants, like the skunk cabbage, display kinds of endothermy.

The point is that these variations are found in living animals. Still other combinations may have occurred in extinct animals, particularly during the time that endothermy as we see it now in birds and mammals may have been in the process of evolution. For instance, a number of paleontologists, notably Nicholas Hotton III, think that for some of the larger dinosaurs, their size alone may have been enough to hold heat once they warmed up, so they could have stayed homeothermic and tachymetabolic without having any internal regulatory mechanism. This is called "mass homeothermy," the mass of the animal keeping the temperature high and steady. In fact, some large mammals, like elephants, because of their size make use of mass homeothermy, willy-nilly. In other words, the retention of heat because of mass is a function of the laws of physics; once a creature or object reaches a certain size, such heat retention is inevitable.

Some paleontologists have also said that because of the dinosaurs' often huge size, endothermy would not have been realistic. It would have produced too much internal heat and the creature's mass would retain it, causing overheating. I don't really follow this argument, because we know of extinct elephants and rhinoceroses from the Miocene era in India that were 18 feet high at the shoulder. This is as big as many dinosaurs. And yet these were mammals. There is no question that they were endotherms.

Among all these various suggestions, my own guess as to how dinosaurs regulated their temperature is probably closest to Bakker's. I suspect that mass homeothermy was important, but not as a complete explanation. If mass homeothermy is all that's going on, it begs the question of what the young did about their temperature regulation. The young (even of sauropods) were too small for their mass to have any significant effect on heat retention. I suspect that most of the dinosaurs were true endotherms, with internal regulatory systems to keep their temperature up high when it was needed. Thus they were able to stay warm and grow rapidly when they were young. And when they grew, the larger ones took advantage of the inevitable effects of mass homeothermy.

But a suspicion is not a scientific demonstration that such is the case. Evidence is needed. Evidence that I think we don't yet have. And the question, the one I faced in Paris, was: What more did we need to show that, beyond a reasonable doubt, at least some dinosaurs were endotherms? One thing was clear: fast-growing bone indicated a high metabolism. If we could show, in a growth series, that the dinosaurs at the Willow Creek anticline had, throughout their lives, a preponderance of the fastest-growing sort of bone, then we would know they had a high metabolism. Furthermore, if they had it at all sizes, it could not have been dependent on just being big enough to hold in heat. But how might other animals have fared in the same environment? What if the environment was hot enough that even the coldblooded animals grew like a house afire? This was a question Ricqlès asked, and he suggested that the answer could be had only if we could find a growth series of an animal that we knew was coldblooded (ectothermic, poikilothermic and bradymetabolic). Well, we didn't have it in Paris, but we did have it. The fossils of the varanid lizards from Egg Mountain showed several stages of growth. Furthermore, we had one other piece of evidence that Ricqlès said was necessary: we had some mammal fossils from Egg Mountain. So we could also see what kind of bone a classic warmblooded animal (endothermic, homeothermic and tachymetabolic) had. We didn't have a mammal growth series, which would have been ideal, but when it comes to fossils you have to work with what you can get.

That, it seemed to me, would be enough. After all, if the lizard showed slow-growing bone and the mammal and the dinosaurs fast-growing bone, from the same environment, then within the limits that you can know anything in science, we would know that those dinosaurs were true endotherms, as warmblooded as birds. Ricqlès disagreed. Something else was missing, he said. We also needed growth series for contemporary animals. One of the things Ricqlès kept pointing out to us was the lack of a well-documented growth series of reptiles, birds and mammals grown in the same environment. Paleontologists and histologists had been talking about the differences among extinct animals and what they showed about their physiology, but nobody had nailed down the differences among living animals as a baseline.

Well, with living animals, unlike fossils, you can set up a project so that you get exactly what you want. At Montana State University we have a veterinary school; we have the facilities for raising animals. In the same environment we are now growing an emu, a cow and a caiman. Two hotbloods and a coldblood. During their growth we are surgically implanting platinum wire bands, wrapping them around the femur at regular stages of growth. When they reach adulthood we'll kill them, make thin sections of their bones, and read, from the space between the wires, how those bones grew.

If we're lucky, when we take the dinosaur, lizard and mammal fossils from the Willow Creek anticline and compare the differences in the way their bones grew to the differences among an emu, a caiman and a cow from the same time and temperature in the Holocene (the age we're living in now), then sit down to analyze the data with Ricqlès, we should come as close as is possible to definitive evidence for how dinosaurs regulated their temperature. I think they were true endotherms, with a high steady temperature and a fast metabolism, and I think that's what the data will show.

One thing I want to mention about temperature regulation, as an afterthought, which I think all paleontologists and I suppose all biologists agree on, is that warmbloodedness, by which I mean the deluxe model (the "warm" side on all three distinctions), is just one metabolic strategy. And it's expensive. It takes a lot of energy to keep your temperature at 98.6 when the outside temperature drops. If you read guides on winter hiking, they'll tell you to take twice as much food as you do in the summer. Essentially you're eating one breakfast to give you the energy to hike and a second breakfast to keep the internal fires going. Food is energy, just as surely as fuel oil or wood that burns in a stove. Most reptiles, in contrast, get away cheap. Since their temperature rises and falls, they just don't need to eat as much. As anyone knows who has heated a big house or driven a gas guzzler during one of the gasoline crises, bigger, faster and hotter is not always better. There are many benefits to going economy style, and this is true in all aspects of life—the biological as well as the automotive. Dinosaurs might be more interesting if they were warmblooded, because it would distinguish them from other reptiles, but it wouldn't mean they were superior.

PERHAPS THIS ACCOUNT makes it seem that I came away from Paris disappointed. It's true that I did not come away with the answers. But I came away with better questions, and better techniques for answering them. It may seem hard to believe, but that's almost as satisfying as the answers themselves. From the outside, science may seem like a collection of answers, a course in "How the World Works." From the inside, it doesn't look like that at all. From the inside, science looks like a bunch of people doing crossword puzzles. It's the doing them that's fun. If you solve one, you don't stop; you look for another.

And from the inside, the way we solve the puzzles, with new approaches, new tools, is easily as important as the solutions. One of the things we brought back from Paris thanks to Ricqlès was some expertise in paleohistology and the chance to bring this way of investigating fossil bones more into the mainstream of paleontology. Not that paleontologists have been unaware of it; but it hasn't been used as much as it should be. After all, our business is figuring out these bones. It seems to me we ought to invest as much time in figuring out their internal structure as we do in figuring out the structure of the rock beds we find them in. Thanks to the Paris trip, Jill made a big contribution in this direction.

In October 1985, at the annual meeting of the Society of Vertebrate Paleontology, Jill Peterson won the prize for the best student address—the Romer Prize, which almost invariably goes to a graduate student. At the time Jill was a sophomore at the University of

Colorado. The reward was not for genius, or for a technically difficult study, although I'm not taking anything away from either her intelligence or her experimental technique. What she had done was to put together thin sections of femur and scapula showing nine stages of *Maiasaura* through the dinosaur's growth from embryo to adult. She made a chart showing these nine stages of growth. Then she took other specimens—of hypsilophodontids, for example—for which the stage of development was known. She compared them to the nine-stage *Maiasaura* series to try to match them. She looked, for example, at the number of vascular canals. What she wanted to do was to come up with an estimate of a sample's developmental age based on how well it fit the *Maiasaura* chart. She then compared these estimates with the actual stage of growth of the sample bone. Invariably, comparison to the *Maiasaura* chart produced an accurate estimate of the developmental stage of the sample. Put another way, with this simple method, using an established growth series, you could take a piece of dinosaur femur, compare it to this series, and find out whether it was a hatchling, or half grown, or an adult.

It doesn't sound so surprising, does it? And yet for years paleontologists have been wondering how to tell adult from juvenile dinosaurs. (Earlier I described the number of misidentified fossils.) They have sometimes gone through extraordinary contortions to get at this knowledge. You don't always have a full bed of bones with various stages of growth to compare. You don't always have a full bone, so that you can look to see the developmental signs that are obvious without a microscope, and those only indicate whether or not the animal was very young when it died. So, what to look for? Some researchers working on hypsilophodontid bones tried various measurements, such as getting a ratio of tibia length to width, that would give the stage of growth for a given bone. This never quite came off. In fact, there was no accepted way of determining the stage of growth of isolated dinosaur fossils.

The chart still needs to be tested to confirm its general value. So far, we've shown that it works for hadrosaurs, hypsilophodontids and ceratopsians that we've found. Of course, even if this specific chart needs to be refined, simply having done it shows the way for other charts, if they are necessary for, say, saurischian carnosaurs. In that sense, even though the histological information has always been there for somebody to use (there are other growth series of fossils available), Jill's study broke new ground. When Jill finished her presentation, a paleontologist next to me grumbled that he was not so impressed. He said that anybody could have done it. I agree with him. Anybody could have. But nobody else did.

CHAPTER 8

BABIES EVERYWHERE

We spent six full seasons at the Willow Creek anticline, from 1979 through 1984. Then the Peebles family decided they wanted to trade in the land on which we'd found the fossils. Because the site was something Montana was proud of, particularly since the fossils and reconstructions were staying in Bozeman, the Peebles hoped to give the state the fossil-ridden badlands in return for a larger amount of good pasture land. I don't quite know the reasoning behind the desire for the swap, except that I suppose with 20 or so crew members on the site each summer and a few hundred visitors, the Peebles must have felt that this small chunk of land had already become somewhat public.

The state of Montana and the Peebles family couldn't come to terms, and this meant the Peebles did not invite us back for the 1985 season. In 1987, the Nature Conservancy bought the land in question. Now, if we want to, we can go back, and we may; however, since we left the anticline, we've found other sites that demand our attention.

In looking back at the dig at the Willow Creek anticline, I'm struck by several things. One is that the dig changed my professional life. I went from preparator to curator. I started getting significant grants. I became able to muster the funding and institutional support to conduct the kind of work I wanted to do. The dig also took charge of the direction of my work. What I mean is that from the finds at the dig, new questions and problems emerged for me to follow. By the time we finished at the anticline, we knew we had hold of a good piece of string that would pull us much deeper into the tangle of dinosaurs. We really had no choice but to keep pulling.

You can summarize the finds at the Willow Creek anticline in terms of quantity, of firsts. In two square miles, we uncovered the first clutches of dinosaur eggs in North America and the first nests of baby dinosaurs anywhere. We found the first evidence of colonial nesting for two quite different species of dinosaurs. We uncovered the first evidence of parental care in dinosaurs—evidence that made *Maiasaura*'s behavior more reminiscent of birds than of reptiles. We found the remnants of the largest herd of dinosaurs yet known. And we discovered two new species of dinosaurs: *Maiasaura peeblesorum* and *Orodromeus makelai*.

This is, however, a very superficial way of looking at the dig and at paleontology. It's more like compiling baseball statistics. What's important is to look at what the finds meant, what they told us about dinosaurs that we hadn't known before, and what they suggested we might be able to find out about dinosaurs if we just kept looking a bit longer.

Certainly, the most important finds were the nesting grounds—the eggs, the embryos, the babies. They were important because we learned things we hadn't known before and we were able to guess at others. We learned without any doubt that at least two species of dinosaurs gathered in colonial nesting grounds, collecting in birdlike flocks to lay their eggs. We learned that both the maiasaurs and the hypsilophodontids grew rapidly, like birds, and that the maiasaur adults seem to have cared for their young by bringing them food in the nest. There are critics who say that we can't be sure of this behavior, that the evidence just isn't sufficient, but I've explained why I can't see any other explanation for what we found. And I think these discoveries alone caused a significant shift in our attitude about dinosaurs, not so much about what they looked like, or even about their physiology (although it seems clear to me that these fossils and the studies we've done of them support the idea of endothermy in dinosaurs), but about the kind of creatures they were, about how they *behaved*.

To envision dinosaurs gathered in aggregations to lay eggs, to see them bringing food to their young, is to imagine them in a new way, to begin to see that dinosaurs were something different. I don't mean to say that the Willow Creek anticline fossils alone are causing this change. But they provide some rigid, documentary skeletons for theory and speculation. Having considered what we found at the anticline, I think it's undeniable that the dinosaurs weren't just big lizards. They weren't mammals or birds, either, and I don't want to anthropomorphize them when I talk about babies; their child care was not the same as the care that goes on in our own species. They were *dinosaurs*, different from all living creatures in many ways, some of which we know about and some of which we'll probably never know about. The picture we have of dinosaurs will always have its blank spots, places where we have to guess to fill in an image, the way we sometimes guess when we reconstruct skeletons.

The fossils from the Willow Creek anticline also reinforced the notion that some dinosaurs lived in herds. This was an idea for which there was already fossil evidence, and the big bone bed we found provided a new and compelling reason to think of some dinosaurs as herd animals. The maiasaurs lived in aggregations of 10,000 individuals or more. And finding the herd led me to speculate that perhaps the dinosaurs exhibited some of the behavioral characteristics of modern herbivorous mammals beyond simple herding. As big herbivores in social groups, they may have been solving similar evolutionary problems. As I said earlier (and as others have said before me) I think the horns and frills, and crests, and vertical plates we see on various dinosaurs did not evolve as means of defense any more than the horns of bighorn sheep evolved as means of defense. My opinion is that they were sexual signals, a means for the males to advertise themselves and compete for females.

Finally, the physical nature of *Maiasaura*, her evolutionary conservatism, and her geographical location in the mountains gave me some very specific ideas about how dinosaurs, particularly the hadrosaurs and lambeosaurs, evolved during the Cretaceous period in North America. To summarize again what I described in Chapter 3, I felt that the movements of the inland sea were driving evolutionary change, compressing habitat and then opening it up wide, in each case causing a burst of evolution and the appearance of many new species. I expected to see the most rapid evolutionary change around the peak of each transgression, when the sea extended its farthest. I thought sediments from those times would provide a lot of new species.

To me the Willow Creek anticline was as much a fountain of ideas as it was a trove of fossils. And in science what you do with ideas, with guesses or hypotheses or tentative conclusions, is to try to confirm or disprove them. Even though paleontology is a historical science, testable predictions are still possible. A paleontologist might say: "Well, look, I know where to find baby dinosaur bones. I've found enough that I figure I've got this puzzle solved. You give me the geologic maps, and I can pick out the spots where you'll find nests, eggs and babies, from the Triassic, the Jurassic, the Cretaceous, from Asia, Europe, South America, North America—anywhere, any geologic time, I can tell you where the baby dinosaurs are." If anyone were to say that, well, the way to prove or disprove the hypothesis would be to take the predictions and go out and try them.

AS THE DIG at the Willow Creek anticline was drawing to an end, all of us, Bob, Jill and I, began to think about where we should go next and what we should look for. The Willow Creek anticline had been wonderfully productive, and it had been a great experience scientifically and personally for most of the people who worked on it. But it had, originally, been the result of blind luck. Marion Brandvold had stumbled onto the baby bones and then we stumbled into her rock shop. We didn't want the story to end with this one find. We didn't want it to be just a happy bit of serendipity that produced a lot of baby bones. We wanted it to be a beginning, to represent a starting point from which we and other paleontologists would go on to find more babies and more nesting grounds, to penetrate more fully the mystery of how the dinosaur babies hatched and grew, of how dinosaurs lived and evolved. We wanted to validate the find by coming up with another like it.

I was convinced that, for starters, if I stuck to the Two Medicine formation, I would find babies. In terms of geological characteristics, I wanted a site like the anticline. The new site had to have, as its predominant sediments, green or red mudstone and calichi, because these were signs of a specific kind of upper coastal plain environment. Given those characteristics, it also had to have abundant bone and eggshell fragments on the surface to make it worth exploring further.

To conduct the search I relied on several things, the first being my memory and the 20 years I'd spent poking around this and that formation in Montana. I knew where a lot of the good exposures of eroding fossil-bearing rock were in the Two Medicine formation, so I knew where to start. In some cases, I had already seen bones or eggshell during one exploration or another. (I do a lot of walking, just looking for possible fossil sites.) I also checked the geologic maps and went over reports of previous paleontological expeditions in the Two Medicine formation that had found juveniles or eggshell. Obviously, such spots would be prime for further digging, since the early paleontologists had often just collected from the surface and moved on. From this search, several candidates for the next dig site emerged.

In 1984 Jill and I explored a spot southwest of Choteau that we called Red Rocks. We found a complete protoceratopsian skeleton, a lot of bones of baby dinosaurs, and a few eggs. I put a crew on this spot in 1985 and they uncovered more, including an adult ceratopsian and hadrosaur. But this was not our only site, nor was it the best one. Also in 1985, after a long period of negotiations, I got permission from the Blackfoot Indian Tribal Council to explore and excavate specimens from tribal lands. These included a spot called Landslide Butte. I'd picked out this site in 1984. I knew Charles Gilmore had found juvenile dinosaurs and eggshell fragments up there. And I'd visited the reservation and seen the site, so I knew it fit the bill. There were visible eggshell fragments, juvenile bones, and red and green mudstones and calichi.

Actually, I first learned of this site in the winter of 1978, before the visit to Marion Brandvold's rock shop. It was the winter after I'd found the lumpy fossil from the Two Medicine formation that turned out to be an egg. When I was trying to be sure about just what I had, I went to the Smithsonian research collection to look for comparable objects. I saw what looked to me like eggshell fragments, with no description attached. I then dug out the field diary of Charles Gilmore, who had found those fossils. He said that he thought they were eggshell fragments, and he described where he found them. That was in 1928, and the location was Landslide Butte. He never published anything on those fragments, even though he found them shortly after the widely publicized discovery of the first dinosaur eggs in Mongolia. If he had, they would have been the first reported eggshell fragments in the Western Hemisphere. Instead the honor went to Glenn Jepsen of Princeton, who in 1930 found fragments in the Hell Creek formation in Montana.

Why Gilmore did not publish that find, I don't know. Nor do I know how he could not have seen the abundance of other fossils at Landslide Butte. Because when we finally got there, the site was so obviously rich that it seems Gilmore should have seen the fossils even if he had galloped through on horseback. After only one season there, we realized that we had found a new site that made the Willow Creek anticline look almost barren. The Willow Creek anticline had been a surprise. Landslide Butte was a shock.

In a total area of a few square miles, we have so far found two large bone beds of ceratopsian dinosaurs and three large bone beds of hadrosaurs and lambeosaurs. We've found a big ceratopsian nesting ground, the first of its kind. And, in the most astonishing discovery, we found a nesting ground of hadrosaurs that is a mile wide, three miles long and three horizons deep. That is to say, stacked on top of one another are three nesting grounds of this one dinosaur, each three miles square. And these nests are far from being sparse or hard to find. At the Willow Creek anticline, for all the bones it yielded, we had to crawl on our hands and knees to find the signs of a nest and we had to train our eyes to look for them. At Landslide Butte, the abundance is almost embarrassing. We already have hundreds and hundreds of baby bones. There are spots where, without even digging, you can literally shovel up the baby bones. Everywhere we look, we find eggs and babies. The babies range from embryonic to about four feet long, the same size range as the Willow Creek anticline.

There are red beds and green beds and calichi in this nesting ground. There are hillsides with the bones just jumping out of them. The eggs range from fragments to squashed eggs to clutches. The site is not as undisturbed as the anticline was. There was some water movement here, so many of the bones have been moved slightly and so far there are no signs of nest structure. But there are identifiable clutches of eggs and there are some skeletons that have all the bones together, so we know the water movement was not too extensive. And so far, just as in the Willow Creek anticline, what we find are baby bones only—no adult bones. Wherever we look, wherever we dig, we find baby bones. At a rough guess there could have been 500 nests in this area—500, that is, on each of the three fossil horizons.

In itself, this is tremendously satisfying. But it is, after all, a site that we went to because I had seen in Gilmore's notes a reference to eggshell fossils and thought I should follow it up. This is upper coastal plain (that's all there was when the sea was that large) but still, in order to say, "Now I know where to find baby dinosaurs anywhere in the world, now the problem is solved," you can't just go to another part of the same coastal plain where you've already found babies. You need a broader pattern. And that is what I've tried to find, at least in outline, to support my hypothesis about where to look for baby dinosaurs.

The hypothesis is simple enough. I think Charles Sternberg was right. He said to look in the uplands, the upper coastal plains. His notion was that the lower coastal plains were too wet and acidic, so that dinosaur eggs would rot or be eaten away. I think he's probably right about the reason. But I *know* he's right about the result.

Both the Willow Creek anticline and Landslide Butte are part of the Two Medicine formation, which is all upper coastal plain. Both of them, in the mudstone sediments I've just described, have eggs and babies. Furthermore, I have looked at other time horizons within the Two Medicine formation and found eggs and babies. For instance, the Red Rocks site, also at the bottom of the Two Medicine formation, yielded eggshell and baby bones.

The Two Medicine formation is not the only one where this is true. The red beds of Mongolia also preserved an upper coastal plain. And I made a few exploratory trips around the American West to look at other formations. I went to a part of the Judith River formation that butts up against the mountains. The terrain there is midway between upper and lower coastal plain. I found a lot of eggshell fragments, and other paleontologists (Mark Goodwin and Peter Dodson) have found more eggshell and juvenile bones. This, by the way, is just west of the spot in the Bear Paw shale where I first looked for baby dinosaurs. If I had gone west to find out where those juvenile fossils had been before they were carried to the sea, I would have ended up there.

I picked, by looking at a geologic map, another part of the Judith River formation that is an upper coastal plain. This one is in an elk habitat in Yellowstone National Park. I found eggshell fragments in the one day I spent there. I went to the Hell Creek formation, which is the latest Cretaceous, just before the dinosaurs went extinct. This was a spot where Glenn Jepsen had reported finding eggshell fragments. I searched for his location to see what it looked like. I knew it was an upper coastal plain, and when I got there I saw just what I expected: red mudstone, with plenty of calichi.

I went to the Wayan formation in Idaho. John Doar from the University of Michigan had found eggshell fragments in this formation in 1983. This is in the lower Cretaceous. I looked at the area he found them in, and again it is an upper coastal plain with the same kinds of mudstones and calichi. Interestingly enough, the lower coastal plain there is dry, not swampy. It has red beds and calichi, too, a similar environment, but no eggshells. I don't know why that is, unless migrating habits were already instinctive in dinosaurs that lived on that lower plain. In any case, it's the upper coastal plain where the eggshells are.

After our work in Paris, Jill and I also went to sites in the south of France, one of the more beautiful environments to look at the sites where what seem to be dinosaur eggs have been found. Here, set in the middle of farms and vineyards, are little badlands the size of postage stamps, again with mudstones and again part of an upper coastal plain.

The find that probably gives me the most satisfaction is not one of my own. For a long time I'd been telling Phil Currie, a good friend and assistant director of the Tyrrell Museum in Drumheller, when he stopped down to visit our sites, that he'd find eggs and nests, too if he'd just go look in his own backyard. You see, the Two Medicine formation extends into Canada. It has traditionally been called by a different name up there (the Oldman Formation), but there's no geological reason for this; the different names are the result of confusing national boundaries with real boundaries.

Well, in the summer of 1987, Phil finally got enough time to go poking around for eggs near the Milk River, in preserved upper coastal plain. After looking for a few weeks, he and his crew found eggs. To be precise, Kevin Aulenbach found the first signs of them. When the crew investigated further, it turned out that there were about seven clutches of hadrosaur eggs. The eggs contained perfectly preserved, articulated dinosaur embryos, something nobody, anywhere, had ever seen before. Phil described the find to the press as the most important paleontological discovery in Canada in 50 years.

It's obvious. The dinosaur eggs and baby dinosaurs are in upper coastal plains. I should emphasize once again that when it comes to dinosaur fossils, the only alternative to an upper coastal plain is a lower coastal plain. All over the world, dinosaur fossils are found in one of these two kinds of deposits. Inland areas, if dinosaurs lived in them, were just not preserved by any kind of deposition. So far, lowland areas have rarely yielded eggs or babies. But the upper coastal plains have yielded and will continue to yield them. The puzzle of baby dinosaurs is solved. They're in the upper coastal plains, just as Sternberg said. I'll say it flat out. I can tell you where to find baby dinosaurs. For any geologic period during the time dinosaurs lived, on any continent on which they lived, go explore an upper coastal plain, find mudstone deposits left by stream overflow, and you'll find dinosaur eggs and babies.

NOW THE BIG QUESTION: If finding these fossils is so easy, why didn't anybody do it before? It's a hard one to answer. Part of the reason lies in the relative abundance of fossils. I've talked a lot about all the discoveries we made at the Willow Creek anticline and what a rich site it was. But I've also described how we crawled on our hands and knees to look for these little bones and eggshell fragments. Except at Landslide Butte, which is a true anomaly, they don't just jump out of the ground at you. It's hard, dry work looking for bones in the Two Medicine formation. You could spend a whole luckless life walking this formation with your head way up high, five or six feet from the ground. If you didn't get down on your hands and knees, you might never find a thing.

In the Judith River formation, on the other hand, a lowland deposit, you literally walk on fossils all the time. Most of it is junk. But it's really tough to leave an abundance of big fossil bones to go prospecting on your hands and knees for these little things that are hard to find. Even after I found an egg in the Two Medicine formation in 1977, I didn't go back, partly for that reason.

Furthermore, after the 1930s, field research on dinosaurs in North America really trailed off. From the 1940s to the '70s, there were not many people looking for dinosaurs, period. Just a few, with little funding, were out prospecting. Now, with more exploration going on, the total amount of money given out by the National Science Foundation to all the vertebrate paleontologists in the country for research on dinosaurs is no more than $500,000 a year.

I can count on my fingers the full-fledged excavations for dinosaur fossils on the North American continent, and half of those are in Alberta, where the government seems to have a soft spot for paleontology. With hardly anybody out looking for baby dinosaurs, it has to be hard to find them. You have to be out looking, you have to know what you're looking for, and you have to be lucky.

As we have been, extraordinarily lucky. But there are different kinds of luck. There's the pure dumb luck that led us into Marion Brandvold's rock shop. And there's the help that you can give to luck by being prepared to recognize it when it comes. You have to know what the baby bones look like. And you have to have the luck of knowing exactly what you're looking for and of having an idea of where to look, the kind of luck Phil Currie had. It's always a treat to find new fossils and it always seems like some kind of gift when it happens, one you're never entirely prepared for. But you do prepare. And sometimes, what Branch Rickey said about baseball is true of paleontology: "Luck is the residue of design."

FIELD CHRONOLOGY

1978. Partial field season.

July 23. Bob Makela and I stop at the Brandvolds' rock shop in Bynum, Montana, where we are shown some small bones that turn out to be very rare fossils of baby dinosaurs. *Late July, early August.* Bob and I collect from the surface of the site on the Peebles ranch and dig up the first nest of baby dinosaurs ever found, as well as an adult skull discovered by the Brandvolds. The state of the baby bones indicates that these young stayed in the nest for some time and were cared for by adults. We name the new dinosaur *Maiasaura*—Good Mother Lizard. *Late summer.* Amy Luthin finds the first sign of a second nest.

1979. First full year of digging. We set up camp on A. B. Guthrie's land on the Teton River.

June, early July. As we find more maiasaur nests, it begins to seem that we have discovered a colonial nesting ground of maiasaurs. *July 2.* We start to work on a new site that the Brandvolds are working, a jumbled collection of bones of adult and young dinosaurs. This is the very first hint of what we will eventually realize is a bone bed containing fossil remnants of a herd of 10,000 maiasaurs. *July 12.* Fran Tannenbaum discovers Egg Mountain, where we find one egg after another. In the years to come, this site will yield several nesting grounds of a small creature that is a variety of hypsilophodontid dinosaur.

1980. We set up camp on the Peebles ranch in the heart of the Willow Creek anticline, where we will stay for the duration of the dig.

June. We find another apparent maiasaur nest in what is now clearly a preserved colonial nesting ground. *July, August.* We work on the Brandvold site and similar deposits of adult and juvenile bones without yet realizing they are connected. We also continue to explore the surface of Egg Mountain, finding more egg clutches and growing more confident that there are other nesting grounds preserved there.

1981. Excavation of the first maiasaur nesting ground is complete. A total of eight nests have been uncovered on one fossil soil surface, indicating that maiasaurs gathered here to lay eggs. The same appears to be true for egg clutches on Egg Mountain laid by the hypsilophodontid that we will eventually name *Orodromeus makelai*. Together, these finds represent the first evidence of colonial nesting in dinosaurs.

June. Wayne Cancro, arriving early, is forced to move his tent because fossils are poking into his back. We begin to work this site, in camp, and call it Camposaur. It is very similar to the Brandvold site in the kinds of bones and their condition. Throughout the summer we find indications of similar deposits. By the end of the summer, rough stratigraphic measurements will show they are all part of one big bone bed. It seems that an entire herd of maiasaurs was destroyed all at once by some natural disaster. *July, August.* We continue to collect from the surface of Egg Mountain but cannot get very deep because of the hardness of the rock.

1982.

July 5. Bob Makela and crew begin using a jackhammer to cut into the hard limestone of Egg Mountain in search of more hypsilophodontid egg clutches. This season we remove close to 30 tons of rock. *July 7.* Phil Currie finds a very heavy concentration of eggshell; on investigation, his find turns out to be a second maiasaur nesting ground.

1983.

June 28. We find the third maiasaur nesting ground at Egg Gulch. *July, August.* We take another 50 tons of rock off Egg Mountain, and we continue to remove fossils from different parts of the big bone bed. *September 14.* At the very end of the season we find Egg Island, another colonial nesting ground of *Orodromeus makelai*. In one of the two clutches, on the surface, are 19 fossilized dinosaur embryos, the first ever found.

1984. We take another 30 tons off Egg Mountain. The completed excavation yields a total of 12 hypsilophodontid egg clutches in three nesting grounds, as well as fossils of plants, lizards and mammals, and two unidentified kinds of eggs—one apparently from a dinosaur, the other of unknown origin.

September. Will Gavin finds an ash layer common to the Brandvold site, Camposaur and similar deposits. All are clearly connected, confirming that we have one big bone bed. It seems that hot gases from a volcanic eruption killed the herd.

1985. During the winter, Jill Peterson and I go to Paris to study under Armand de Ricqlès and learn more about the tissue of fossilized dinosaur bone. The hope is to develop positive evidence that dinosaurs were warmblooded. The next summer we will begin the search, at new sites, for more baby dinosaurs.

NOTES AND REFERENCES

Chapter 1: Looking for Babies

1. B. Brown and E. M. Schlaikjer, "The Structure and Relationships of *Protoceratops*," *Annals of the New York Academy of Sciences* XL (3) (1940):133–266.

   P. Dodson, "Taxonomic Implications of Relative Growth in Lambeosaurine Hadrosaurs," *Systematic Zoology* 24 (1975):37–54.

2. C. M. Sternberg, "A Juvenile Hadrosaur from the Oldman Formation of Alberta," *National Museum of Canada Bulletin* 136 (1955):120–122.

3. M. M. Chow, "Notes on the Late Cretaceous Dinosaurian Remains and the Fossil Eggs from Laiyang Shantung," *Bulletin, Geological Society of China* 31 (1951): 89–96.

   W. Granger, "The Story of the Dinosaur Eggs," *Natural History* 38 (1936):21–25.

   G. L. Jepsen, "Dinosaur Eggshell Fragments from Montana," *Science* 73 (1931):12–13.

   V. Van Straelen, "Les oeufs de reptiles fossiles," *Palaeobiologica* 1 (1928):296–312.

   Roy Chapman Andrews, "Where the Dinosaur Hid Its Eggs," *Asia* (January 1924):163–166.

4. R. Gradzinski, "Sedimentation of Dinosaur-Bearing Upper Cretaceous Deposits of the Nemegt Basin, Gobi Desert," *Palaeontologia Polanica* 21 (1969):147–229.

5. We also took with us to the rock shop and the fossil fish dig Gay Vostries, the

manager of the paleontology collection at the Philadelphia Academy of Natural Sciences, who was in Montana that summer to visit several paleontological digs.

6. J. C. Lorenz, "Sedimentary and Tectonic History of the Two Medicine Formation, Late Cretaceous (Companion), North-western Montana" (Ph.D. thesis, Princeton University, 1981).

   —— and W. Gavin, "Geology of the Two Medicine Formation and the Sedimentology of a Dinosaur Nesting Ground," *Montana Geological Society Field Conference Guidebook* (1984):175–186.

7. P. Dodson, "Taxonomic Implications of Relative Growth in Lambeosaurine Hadrosaurs," *Systematic Zoology* 24 (1975):37–54.

8. From C. M. Sternberg, "A Juvenile Hadrosaur from the Oldman Formation of Alberta," *National Museum of Canada Bulletin* 136 (1955):120–122.

   Other reports and comments on this same subject include:

   K. Carpenter, "Baby Dinosaurs from the Late Cretaceous Lance and Hell Creek Formations and a Description of a New Species of Theropod," *Contributions to Geology, University of Wyoming* 20(2) (1982):123–134.

   G. L. Jepsen, "Riddles of the Terrible Lizards," *American Scientist* 52 (1964): 227–246.

9. C. W. Gilmore, "Hunting Dinosaurs in Montana," *Explorations and Field Work of the Smithsonian Institution in 1928* (1929):7–12.

Chapter 2: The First Nest

1. J. R. Horner and R. Makela, "Nest of Juveniles Provides Evidence of Family Structure Among Dinosaurs," *Nature* 282 (1979):296–298.

Chapter 3: The Good Mother Lizard

1. Sources that provided information on dinosaurs, their environment and history include:

   Robert T. Bakker, *The Dinosaur Heresies* (New York: William Morrow and Company, Inc., 1986).

   Edwin H. Colbert, *Dinosaurs, Their Discovery and Their World* (New York: E. P. Dutton & Co., Inc., 1961).

   ——, *Dinosaurs: An Illustrated History* (Maplewood, N.J.: Hammond, Inc., 1983).

   David Dineley, *Earth's Voyage Through Time* (New York: Alfred A. Knopf, 1974).

   Adrian J. Desmond, *The Hot Blooded Dinosaurs: A Revolution in Paleontology* (New York: The Dial Press/James Wade, 1976).

   Url Lanham, *The Bone Hunters* (New York: Columbia University Press, 1973).

   George Gaylord Simpson, *Life of the Past: An Introduction to Paleontology* (New Haven, Conn.: Yale University Press, 1953).

   John Noble Wilford, *The Riddle of the Dinosaur* (New York: Alfred A. Knopf, 1985).

The Diagram Group, *A Field Guide to Dinosaurs* (New York: Avon, 1983).

Dale A. Russell, *A Vanished World. The Dinosaurs of Western Canada* (Ottawa: National Museums of Canada, 1977).

2. R. S. Lull and N. E. Wright, "Hadrosaurian Dinosaurs of North America," Geological Society of America, Special Paper 40 (1942):1-242.
3. J. R. Horner, "A New Hadrosaur (Reptilia; Ornithischia) from the Upper Cretaceous Judith River Formation of Montana," *Journal of Vertebrate Paleontology* (in press).
4. Url Lanham, *The Bone Hunters* (New York: Columbia University Press, 1973).
5. D. B. Weishampel, "Evolution of Jaw Mechanisms in Ornithopod Dinosaurs," *Advances in Anatomy, Embryology and Cell Biology* 87 (1984):1–109.
6. *Ibid.*
7. T. Maryanska and H. Osmolska, "Some Implications of Hadrosaurian Postcranial Anatomy," *Acta Paleontologica Polonica* 28 (1983):205–207.
8. J. R. Horner and R. Makela, "Nest of Juveniles Provides Evidence of Family Structure Among Dinosaurs," *Nature* 282 (1979):296–298.
9. J. R. Horner, "Cranial Osteology and Morphology of the Type Specimen of *Maiasaura Peeblesorum* (Ornithischia:Hadrosauridae), with Discussion of Its Phylogenetic Position," *Journal of Vertebrate Paleontology* 3(1) (1983):29–38.

———, "Three Ecologically Distinct Vertebrate Faunal Communities from the Late Cretaceous Two Medicine Formation of Montana, with Discussion of Evolutionary Pressures Induced by Interior Seaway Fluctuations," *Montana Geological Society Field Conference Guidebook* (1984):299-303.

Chapter 4: Nesting in Colonies

1. J. C. Lorenz and W. Gavin, "Geology of the Two Medicine Formation and the Sedimentology of a Dinosaur Nesting Ground," *Montana Geological Society Field Conference Guidebook* (1984): 175–186.
2. G. J. Retallack, "Fossil Soils: Indications of Ancient Terrestrial Environments," in *Paleobotany, Paleoecology and Evolution*, Vol. 1, ed. K. J. Niklas (New York: Praeger, 1981), pp. 55–102.
3. W. M. B. Gavin, "A Paleoenvironmental Reconstruction of the Cretaceous Willow Creek Anticline Dinosaur Nesting Locality: North Central Montana" (master's thesis, Department of Earth Sciences, Montana State University, 1986).

Chapter 5: The Herd

1. R. T. Bird, "Did Brontosaurus Ever Walk on Land?" *Natural History* 53 (1944): 61–67.
2. P. J. Currie and P. Dodson, "Mass Death of a Herd of Ceratopsian Dinosaurs," in

Third Symposium on Mesozoic Terrestrial Ecosystems, Short Papers, eds. W. E. Reif and F. Westphal (Tubingen: ATTEMPTO Verlag, 1984), pp. 61–66.

D. B. Norman, "A Mass-Accumulation of Vertebrates from the Lower Cretaceous of Nehden (Sauerland), West Germany," *Proceedings, Royal Society of London* B 230 (1987):215–255.

J. H. Ostrom, "Were Some Dinosaurs Gregarious?" *Palaeogeography, Palaeoclimatology, Palaeoecology* 11 (1972):287–301.

F. Von Huene, "Lebensbild des Saurischier-Vorkommens in obersten Keuper von Trossingen in Württemberg," *Palaeobiologica* 1 (1928):103–116.

3. O. Abel, "Die Neuen Dinosaurierfunde in der Oberkreide Canadas," *Jahrag. Naturwiss. Berlin* 12 (1924):709–716.

J. O. Farlow and P. Dodson, "The Behavioral Significance of Frill and Horn Morphology in Ceratopsian Dinosaurs," *Evolution* 29 (1974):353–361.

J. A. Hopson, "The Evolution of Cranial Display Structures in Hadrosaurian Dinosaurs," *Paleobiology* 1(1) (1975):21–43.

R. E. Molnar, "Analogies in the Evolution of Combat and Display Structures in Ornithopods and Ungulates," *Evolutionary Theory* 3 (1977):165–190.

Chapter 6: Egg Mountain

1. Philip J. Currie, "Bird-Like Characteristics of the Jaws and Teeth of Troödontid Theropods (Dinosauria, Saurischia)," *Journal of Vertebrate Paleontology* 7(1) (March 1987):72–81.

2. J. R. Horner and D. B. Weishampel, "A Comparative Embryological Study of Two Ornithischian Dinosaurs," *Nature* 332 (1988):256–257.

3. J. R. Horner, "Ecologic and Behavioral Implications Derived from a Dinosaur Nesting Site," in *Dinosaurs Past and Present,* Vol. 2, eds. S. J. Czerkas and E. C. Olson (Natural History Museum, Los Angeles County/University of Washington Press, Seattle, 1987), pp. 51–63.

4. J. R. Horner and D. B. Weishampel, "A Comparative Embryological Study of Two Ornithischian Dinosaurs," *Nature* 332 (1988):256–257.

Chapter 7: Haute Bones

1. P. Taquet, "Cuvier-Buckland-Mantell and the Dinosaurs," in *Actes du Symposium Paleontologigue,* eds. E. Buffetaut, J. M. Mazin and E. Salmon (Montbéliard: G. Cuvier, 1983), pp. 476–494.

2. The question of who is first on the issue of warmblooded dinosaurs is just too tangled for me to unravel, given the different emphases by different scientists at different times. For anyone who wants to try to sort it out, among the early important papers making reference to this subject are:

R. T. Bakker, "The Superiority of Dinosaurs," *Discovery* (Peabody Museum, Yale University) 3(1) (1968):11–22.

——, "Dinosaur Physiology and the Origin of Mammals," *Evolution* 25 (1971):636–658.

——, "Anatomical and Ecological Evidence of Endothermy in Dinosaurs," *Nature* 238 (1972):81–85.

J. H. Ostrom, "Terrestrial Vertebrates as Indicators of Mesozoic Climates," *North American Paleontological Convention Proceedings* (Chicago, 1969) D (1970): 347–376.

A. J. de Ricqlès, "Evolution of Endothermy: Histological Evidence," *Evolutionary Theory* 1 (1974):51–80.

3. The most valuable recent volume on the issue of warmbloodedness in dinosaurs is *A Cold Look at the Warm-Blooded Dinosaurs* (1980), in the American Association for the Advancement of Science Selected Symposia Series, published by Westview Press, Boulder, Colorado, for the AAAS. Articles of particular use in the preparation of this chapter were:

   R. T. Bakker, "Dinosaur Heresy—Dinosaur Renaissance: Why We Need Endothermic Archosaurs for a Comprehensive Theory of Bioenergetic Evolution."

   Neil Greenberg, "Physiological and Behavioral Thermoregulation in Living Reptiles."

   Nicholas Hotton III, "An Alternative to Dinosaur Endothermy: The Happy Wanderers."

   J. H. Ostrom, "The Evidence for Endothermy in Dinosaurs."

   P. J. Regal and C. Gans, "The Revolution in Thermal Physiology: Implications for Dinosaurs."

   A. J. de Ricqlès, "Tissue Structures of Dinosaur Bone: Functional Significance and Possible Relation to Dinosaur Physiology."

4. D. H. Enlow and S. O. Brown, "A Comparative Histological Study of Fossil and Recent Bone Tissues. Part II," *Texas Journal of Science* 9 (1957):186–214.

INDEX

Index